Routledge Revivals

China's Forest

Forestry and Forest Policy are key issues for the protection of China's natural environment and for its continued economic development. Originally published in 2003, the contributors to this title review the successes of China's forest policies and the growth of its forests over the past quarter-century and examine the challenges facing China's forests and rural environment. *China's Forests: Global Lessons from Market Reforms* is a valuable resource for students interested in environmental studies, international forest policy, and the modern development of China.

China's Forests

Global Lessons from Market Reforms

Edited by
William F. Hyde, Brian Belcher, and Jintao Xu

RFF PRESS
RESOURCES FOR THE FUTURE

First published in 2003
by Resources for the Future, Inc.

This edition first published in 2016 by Routledge
2 Park Square, Milton Park, Abingdon, Oxon, OX14 4RN
and by Routledge
711 Third Avenue, New York, NY 10017

Routledge is an imprint of the Taylor & Francis Group, an informa business

Publisher's Note
The publisher has gone to great lengths to ensure the quality of this reprint but points out that some imperfections in the original copies may be apparent.

Disclaimer
The publisher has made every effort to trace copyright holders and welcomes correspondence from those they have been unable to contact.

A Library of Congress record exists under LC control number: 2003018032

ISBN 13: 978-1-138-94623-1 (hbk)
ISBN 13: 978-1-315-67092-8 (ebk)
ISBN 13: 978-1-138-94627-9 (pbk)

China's Forests

Global Lessons from Market Reforms

EDITED BY

William F. Hyde

Brian Belcher

Jintao Xu

RESOURCES FOR THE FUTURE
Washington, DC, USA

CENTER FOR INTERNATIONAL
FORESTRY RESEARCH
Bogor, Indonesia

Printed in the United States of America

An RFF Press book
Published by Resources for the Future
1616 P Street, NW
Washington, DC 20036–1400
USA
www.rffpress.org

A copublication of Resources for the Future (www.rff.org) and the Center for International Forestry Research (www.cifor.org).

Library of Congress Cataloging-in-Publication Data

China's forests : global lessons from market reforms / edited by William F. Hyde, Brian Belcher, and Jintao Xu.
p. cm.
Includes index.
ISBN 1-891853-67-8 (alk. paper) — ISBN 1-891853-66-X (pbk. : alk. paper)
1. Forest policy—China. 2. Forest management—China. 3. Forests and forestry—Economic aspects—China. I. Hyde, William F. II. Xu, Jintao. III. Belcher, Brian M. (Brian Murray).
SD643.C45 2003
333.75'0951—dc22

2003018032

f e d c b a

The paper in this book meets the guidelines for permanence and durability of the Committee on Production Guidelines for Book Longevity of the Council on Library Resources.

The text of this book was designed and typeset by Betsy Kulamer in Bembo and Gill Sans. The cover was designed by Rosenbohm Graphic Design.

ISBN 1–891853–67–8 (cloth) ISBN 1–891853–66–X (paper)

About Resources for the Future *and* RFF Press

RESOURCES FOR THE FUTURE (RFF) improves environmental and natural resource policymaking worldwide through independent social science research of the highest caliber. Founded in 1952, RFF pioneered the application of economics as a tool to develop more effective policy about the use and conservation of natural resources. Its scholars continue to employ social science methods to analyze critical issues concerning pollution control, energy policy, land and water use, hazardous waste, climate change, biodiversity, and the environmental challenges of developing countries.

RFF PRESS supports the mission of RFF by publishing book-length works that present a broad range of approaches to the study of natural resources and the environment. Its authors and editors include RFF staff, researchers from the larger academic and policy communities, and journalists. Audiences for RFF publications include all of the participants in the policymaking process—scholars, the media, advocacy groups, nongovernment organizations, professionals in business and government, and the general public.

About the Center for International Forestry Research

THE CENTER FOR INTERNATIONAL FORESTRY RESEARCH (CIFOR) was established in 1993 as part of the Consultative Group on International Agricultural Research (CGIAR) in response to global concerns about the social, environmental, and economic consequences of forest loss and degradation. CIFOR research produces knowledge and methods needed to improve the well-being of forest-dependent people and to help tropical countries manage their forests wisely for sustained benefits. This research is done in more than two dozen countries, in cooperation with numerous partners. Since it was founded, CIFOR has also played a central role in influencing global and national forestry policies.

Contents

Contributors

William Hyde's career in forest economics and policy spans 30 years and includes terms with the U.S. Forest Service, Duke University, and Resources for the Future. He is currently affiliated with the Center for International Forestry Research in Indonesia and with the Göteborg University in Sweden. He has been a consultant to more than two dozen organizations in more than 30 countries. His books include *Timber Supply and Land Allocation*, *Forestry Sector Intervention* (with Roy Boyd), *The Benefits of Forestry Research* (with David Newman and Barry Seldon), and the *Economics of Forestry and Rural Development: Empirical Evidence from Asia* (with Gregory Amacher).

Brian Belcher is a Senior Scientist at the Center for International Forestry Research, specializing in natural resources and livelihoods issues. He has had a series of research projects in China focusing on the economics and policy aspects of the bamboo sector in China. He is leading an international comparative analysis of cases of nontimber forest product commercialization and continuing research on the changing role and potential of forests in economic development in Indonesia. He is the coeditor of *Carving out a Future: Tropical Forests, Livelihoods and the International Woodcarving Trade.*

Jintao Xu is Deputy Director of the Center for Chinese Agricultural Policy, at the Chinese Academy of Sciences. His research and writing interests include natural resource modeling, the Chinese paper industry, deforestation, and rural forest reform.

Vince Benziger was Social Science Research Associate, Asia Pacific Research Center, Stanford University, at the time his chapter was prepared. He is currently a consultant in Champaign, Illinois.

Xuhe Chen is Deputy Director General, International Network for Bamboo and Rattan, and Professor, Chinese Academy of Forestry.

David Edmunds was Research Associate, Center for International Forestry Research, at the time his chapter was prepared. He currently works with Red Nation Consulting.

Maoyi Fu is Professor and Principal Scientist, Research Institute of Subtropical Forestry, Chinese Academy of Forestry, Fuyang, Zhejiang, China.

Jikun Huang is Director, Center for Chinese Agricultural Policy, Chinese Academy of Sciences, Beijing.

Shashi Kant is Associate Professor, University of Toronto.

Jari Kuuluvainen is Professor, University of Helsinki.

Natasha Landell-Mills is Senior Research Associate, International Institute of Economic Development.

Dachang Liu is Academic Director, Mekong Institute, Khon Kaen University, Thailand, and Associate Professor, Southwest Forestry College, Kunming, China.

Jinlong Liu is Associate Professor, Chinese Academy of Forestry.

Scott Rozelle is Professor, Department of Agricultural and Resource Economics, University of California, Davis.

Manuel Ruiz Pérez is Associate Professor, Autonomous University of Madrid, and Senior Research Associate, Center for International Forestry Research.

Jeffrey A. Sayer was Director General, Center for International Forestry Research, at the time his chapter was prepared. He is curently Senior Associate, World Wide Fund for Nature–China and Professor of International Nature Conservation, University of Utrecht.

Changjin Sun is Director, Research Center for Ecological and Environmental Economics, Chinese Academy of Social Sciences.

Jussi Uusivuori is Academy Researcher, University of Helsinki.

Xiaosheng Yang is Associate Professor, Research Institute of Subtropical Forestry, Chinese Academy of Forestry, Fuyang.

Runsheng Yin is Associate Professor, Michigan State University.

Daowei Zhang is Professor, Auburn University.

Jiurong Zhang is Professor and Vice-President, Chinese Academy of Forestry.

Yaoqi Zhang was doctoral student at the University of Helsinki and postdoctoral fellow at the University of Toronto at the time his chapter was written. He is currently Assistant Professor, Auburn University.

Foreword

Uma Lele

RECOGNITION OF THE IMPORTANCE of forestry is as ancient in China as it is modern. Zhang Juncheng in his classic "Introduction to China's Forestry Tradition" stressed the recognition of the role of forests in Zhou Dynasty established in 1066 B.C.

The same is true in the western tradition. Forestry's critical role was recognized as early as when Solomon depleted the cedar forests of Lebanon to build his great temple.

Forests are even more important today—not just for their commercial value, but for the income and livelihood they provide to the millions of poor households globally. The environmental degradation associated with the population explosion and the income growth worldwide in the twentieth century has also enhanced their role as a source of environmental values, namely, as habitats for biodiversity, as a store of carbon to mitigate global climate change, as a reliable source of clean water, and in the protection of soils against erosion and the consequent loss of agricultural productivity downstream.

China's farmers demonstrated their own appreciation of the importance of both the market and nonmarket values of forests in the early 1980s as they undertook an unprecedented expansion of tree planting on the lands under their own management. Forest cover on the collective forest and farm lands increased by millions of hectares, as much as 60% in 15 years, a global record in small-scale tree planting.

Despite this period of tremendous forest recovery, China's forests—and particularly its state-owned natural forests—today remain a major environmental management challenge both from a national and global perspective, the latter due to China's sheer size and hence role in the state of the global environment. To better understand the issues of global forest management, China's experience was included as one of the six country cases reviewed by the World Bank as part of the evaluation of its own forest strategy. China's forests were also the focus of a major assessment by the Forests and Grasslands Task Force of the China Council for International Cooperation on Environment and Development published in 2002.

China's Forests: Global Lessons from Market Reforms and the work of its editors and contributors is a comprehensive examination of the contemporary issues facing China's forestry. The book's contributors are distinguished forestry experts with a wealth of experience both in China and the rest of the world. Through thorough analysis, collectively they provide important perspectives on the impacts of China's policies on its forests. They stress China's forestry successes since the beginning of reforms in 1978 and draw lessons from the experience both for China and for global forestry. They identify the challenges that remain and their implications for future policy.

I have no doubt that this volume will be a landmark in the development of thinking about forest policy, both in China and globally.

UMA LELE
Leader, the World Bank's Forest Strategy Evaluation;
Co-Chair, China Taskforce on Forests and Grasslands;
and Senior Advisor, Operations Evaluation Department, the World Bank

Foreword

Genchang Chen

FORESTRY AND FOREST POLICY are key issues for the protection of China's natural environment and, indeed, for our continued economic development. Premier Zhu Rongji identified sustainable development as one of China's two fundamental strategies for the twenty-first century. President Jiang Zemin further stressed the importance of environmental protection at the annual workshop on Population, Resource and Environment on March 13, 1999. Sustainable forestry is a crucial part of this focus on the environment.

The distinguished contributors to *China's Forests: Global Lessons from Market Reforms* have completed an important task—reviewing the successes of our forest policies and the growth of our forests over the past quarter-century. They have also anticipated many of the tasks ahead, for the future of our forest policy and for the management of our rural environment. Our farm foresters have performed well. The new challenges are the development of a less burdensome system of forest taxes, the role of forests in western regional development, and the introduction of a contract responsibility system for state forest management. The contributors to this book have identified these challenges, and they have demonstrated how rigorous quantitative analysis can distinguish among the merits of alternative policies for addressing these issues. Their book will be a valuable asset as we consider our future opportunities.

GENCHANG CHEN
Director General
Department of Policy and Legislation
State Forestry Administration (China)

Preface

FORESTRY PLAYS A GREATER ROLE in the global policy discussion than at any point in recent memory. The most general themes in modern forest policy have to do with sustainable management and, particularly, the institutional arrangements for managing forest land. These themes are often interchangeable with the idea of decentralization from state agency management to management by local institutions, and the successes of both sets of institutions are judged on their ability to produce sustainable flows of market and nonmarket forest products and environmental services—timber and fuelwood; fruits, nuts, forage, fodder, latex, and traditional medicines; and watershed services, natural biodiversity, and carbon sequestration to deter global climate change.

No country's forest policy has changed more dramatically than China's, and many of the changes in China's policies are part of its broader trend toward more decentralized management in all sectors of the economy. China's experience of great and rapid change begs reflection—with the expectation that this experience will be instructive for the continued progress of China's forestry sector and for those in the rest of the world who also struggle with forest policy reform in general and the decentralization of forest management in particular.

A group of research scholars, project managers, and government officials recognized these points, and they agreed on the merits of reflection. They met in June 2001 in Dujiangyan, Sichuan Province, China, to review and synthesize research observations on the impacts of China's rural, industrial, and financial reforms on the forest sector. The antecedent literature was sparse, and most of it focused on the evolution of property rights. This was the central feature of China's rural reforms, and it continues to be an important feature of its forestry reforms, even at the beginning of the twenty-first century. The group that met in Dujiangyan agree on this, but they also recognize the increasing demands on the forest arising from an extended period of general economic growth and they perceived emerging concerns for both the nonmarket environmental services of forests and increasing regional disparities in rural household incomes. Many of the group have continued to meet and exchange insights on this range of topics. *China's Forests: Global Lessons from Market Reforms* reports the fundamental insights emerging from their discussions.

China's reforms either originated in or targeted a number of different sectors. Therefore, it has been important for our group to involve participants from a range of sectors. The original symposium in Dujiangyan was entitled "Policy Reform and Forestry in China: Lessons for China and the World." It was co-organized by the Center for International Forestry Research (CIFOR), China's State Forestry Administration, the Research Center for Ecological and Environmental Economics of the Chinese Academy of Social Sciences, the Chinese Academy of Forestry, the Center for Chinese Agricultural Policy of the Chinese Academy of Sciences, and the Sichuan Forestry Department. The participants themselves included national and international forest and agricultural policy researchers and research managers, policymakers from different branches and levels of government, practitioners from various development projects, and representatives from several other countries in the region. Representatives of donor and development agencies also attended.

The symposium was organized around a series of themes: history and the grand themes of the reform; decentralization; prices, taxation and regulation; investments in forestry; extrasectoral policy impacts on forestry; impacts on poverty and rural households; environmental impacts. The discussion of each theme began with one or more primary papers that were selected for their recognized research or analytical basis. (The organizers wanted to avoid "opinion pieces.") Key discussants synthesized the basic points of these papers, identified remaining unanswered questions and emerging issues, and directed a plenary discussion on each theme. The general discussion went beyond the boundaries of "forestry" to consider a wide set of factors that interact in a very dynamic environment to influence resource management and development.

The chapters of this book emerged from some of the papers presented during the symposium and the discussion that continues to develop. In fact, the International Institute for Environment and Development, Forest Trends, and the Forest and Grasslands Task Force of the China Council for International Cooperation on Environment and Development have each held their own symposiums on China's forest policies since we began our discussion. Those symposiums provided further opportunity for many of our participants to refine their insights in preparation for this book.

Acknowledgements

We are grateful to the co-organizers of the symposium and to our sponsors throughout the effort to put the book together: the Ford Foundation, Australian Centre for International Agricultural Research, the International Development Research Centre, and World Wide Fund for Nature–China. CIFOR itself provided core support. Our steering committee included Jeffrey Sayer, CIFOR, now World Wide Fund for Nature–China; Jin Puchun, International Forestry Cooperation Center, State Forest Administration; Zhang Jiurong, Chinese Academy of Forestry; Changjin Sun, Research Center for Ecological and Environmental Economics; Jikun Huang, Center for Chinese Agricultural Policy, Chinese Academy of Sciences, and Brian Belcher, CIFOR. William Hyde, University of British Columbia and Göteborg University; Liu Dachang, CIFOR and Southwest Forestry College in Kunming; David Edmunds, CIFOR; and Hu Yuanhui, International Forestry Cooperation Center, State Forest Administration, provided support. In addition, we must give special thanks to Professor Runsheng Yin of Michigan State University for the original suggestion; to Manuel Ruiz Perez of CIFOR and the Autonomous University of Madrid, who recognized its merit and supported its initial development; to Liu Dachang for his close collaboration throughout our effort; and Don Reisman of Resources for the Future, who encouraged us to turn our insights into an organized book.

We received special encouragement, thoughtful critique, and other useful advice and assistance from a number of others over the course of the two years it has taken us to prepare the symposium and the book: Ian Bevege and John Fryer, Australian Centre for International Agricultural Research; John Graham and Stephan McGurk, International Development Research Centre; Zhu Chunquan and Jim Harkness, World Wide Fund for Nature–China; Sen Wang, University of British Columbia; Hein Malee, Ford Foundation; and Mafa Chipeta, formerly with CIFOR and now with the U.N. Food and Agriculture Organization. Titin Suhartini and Rita Mustikasari of CIFOR assisted with organization and correspondence of all sorts.

Finally, we would like to note our own convention of using western order (given name first) for the Chinese names in this volume. A Chinese-language version of the book will use Chinese names in their standard order.

We appreciate the assistance and encouragement of each of these individuals and institutions. We hope that our own effort has been up to their expectations and that this book will serve as a key reference on forest management, policy reform, and forest development for China and for the rest of the world.

William F. Hyde
Brian Belcher
Jintao Xu

Note: China, showing six forest regions and the provinces and autonomous regions within each. The North region includes Hebei, Neimenggu, Shanxi, and Tianjin. The Northeast region includes Heilongjiang, Jilin, and Liaoning. The Northwest region includes Gansu, Ningxia, Qinghai, Shaanxi, and Xinjiang. The South Central region includes Guangdong, Guangxi, Hainan, Henan, Hubei, and Hunan. The Southeast region includes Anhui, Fujian, Jiangsu, Jiangxi, Shandong, and Zhejiang. The Southwest region includes Guizhou, Sichuan, Xizang, and Yunnan.

CHAPTER 1

Introduction

William F. Hyde, Jintao Xu, and Brian Belcher

CHINA HAS UNDERGONE a remarkable transition since the introduction of rural reforms in 1978. The summary effects are well-known: average annual growth in gross domestic product (GDP) regularly exceeding 10%, and more than a quadrupling of per capita income. Rural households were the first to benefit. As industrial, financial, and then trade reforms followed the initial rural reforms, nearly all members of Chinese society have benefited.

The reforms have continued since 1978—gradually, but with dramatic results—and China's leaders now seek to maintain the growth that the populace has come to expect. However, with reform and growth, new problems have arisen. China avoided most of the trials suffered by its neighbors as a result of the East Asian financial crisis of 1997, but 20 years of growth have created regional income disparities and new environmental problems, including some with serious implications for forest management.

This book is about the impacts of China's reforms on the forestry sector of its economy. Forestry-sector reforms followed agricultural reforms in time and in structure, but within the forestry sector, neither the marketing nor the administrative reforms have been as complete as they are in agriculture. Moreover, forestry's longer production period delays some of the evidence of improved performance. For these reasons, it is not surprising that the impacts of reform on forestry, while substantial, have not been as immediately and uniformly spectacular as those for agriculture. For example, the value of agricultural output has increased fourfold, while China's forest cover has only doubled and the level of annual timber harvests has remained largely unchanged since 1978 (*China Statistical Yearbook* 2000).

Forestry is also different from agriculture in that spillover effects from reforms in other sectors had a profound effect. By some measures, such as the number and area of trees planted, these spillovers may have had a greater impact on forestry than some of the new forest policies themselves. In contrast, we would argue that spillover effects have gone the other direction for agriculture, from agricultural policy to benefit of the rest of China's economy.

Finally, the forestry sector provides an interesting perspective on China's emerging distributive and environmental issues. In many societies, forest-based communities and forest-dependent households tend to be poorer than communities and households in other sectors of the economy, and their income flows tend to be more variable. These distributive issues may be especially important in China, where a historically egalitarian society increasingly objects to growing disparities in regional development and household wealth. Forest ministries, international donors, and nongovernmental organizations often look to the forestry sector for activities that can modify these disparities, but the limits to the sector's potential in this regard are not well-established, in China or anywhere else.

And what about the environment? Forests are conservatories of a great many environmental values. In China, many of these values have increased with the improvements in household wealth, and China's population can now afford a more farsighted view of its general welfare. The net environmental effects of economic growth, however, are complex. Total forest cover has expanded, and this has helped contain the effects of wind erosion and desertification, two problems whose combined effects may be worse in China than anywhere else in the world. Meanwhile, China's natural forest cover and its reservoir of biodiversity have declined as a result of timber harvesting and land conversion to agricultural crops and for use by grazing livestock. Furthermore, industrial growth has been a source of pollution. The pulp and paper industry, an important component of the forestry sector, has become the largest polluter of China's rural environment. Understanding the complexity of effects on the forest environment as well as the inducements for them that have been generated by China's reforms is surely a crucial step toward improving China's long-term welfare.

Objective

The objective of this book is to examine the experience of China's forestry sector with respect to the full set of forest and nonforest policy reforms introduced since 1978. The authors of the chapters rely on empirical evidence when it is available and tie specific forestry-sector impacts to select categories of policy inducements when these ties can be identified. They review what has been established in the analytical literature of the field and add insights from their own research. They will conclude that some old questions remain unanswered even as they identify newly emerging policy issues.

The lessons of this book are important for China, and for the rest the world as well. For China, the analyses in each chapter provide rigorous evidence on those policies that have worked and on how well they have worked. Furthermore, they provide some support for, as well as some contrast with, the agricultural benchmark that has dominated the prior analytical literature on China's rural reforms. Agricultural reforms developed from a diversity of local experiments. Forestry reforms have added still more diversity in the terms of contracts for land use; in tax systems, marketing policy, and administration; and in the emerging income distributive issues.

The unanswered questions and emerging issues that arise from these analyses will help us recognize the unanticipated effects of China's reforms. These should be important ingredients in the decisionmaking process for future reforms as China's government continues to pursue economic growth and broadly improved welfare of all its people, from the 3% of its population employed in the forestry sector to the much larger share that consumes forest products and benefits from forest-based environmental services.

In 2003, China has just become a member of the World Trade Organization. It is embarking on a new western regional development program and a forest protection program as well. All three promise serious implications for the future of forestry. Domestic timber harvest levels have already declined, and roundwood imports have increased sharply (182%, or from 4.8 million to 13.6 million m^3 in the first two years of these programs). Yet many of the effects of China's newest government programs remain matters of debate. The analyses of prior reforms in this book may help us understand the right questions to ask about the new programs. They may even anticipate the effects of newer reforms and surely they provide material for helpful reflection as China's leaders consider improvements in the design of long-term forest policy.

The lessons from China are also important for the rest of the world. For example, many eastern European and some Asian countries are undergoing a similar economic transition from centrally planned to more market-oriented economies. Others, such as Indonesia and the Philippines, are making the transition from economies with highly concentrated political and market power to a wider distribution of economic authority and less-administered markets. Still others, such as India, are attempting broad market deregulation. All can learn from China's experience.

China's experience is even more broadly relevant because so many of its forests—like those of most countries, developing and developed—have been under the authority of a central forest ministry. China, like most countries today, faces the difficulties of centralized forest management in the presence of powerful local values. In fact, the decentralization of forest management and the development of management systems that are more responsive to local economic and social pressures may be the most global issue of modern forest policy. Some degree of decentralization is known as "community participation" in the United States and Canada; "joint forest management" in India; and "community forestry" or "community-based resource management" in Bolivia, Colombia, Mexico, Nepal, the Philippines, Tanzania, and Zambia. Similar terms are used for related experiences in a host of other countries as well. China has had more experience than any other country in transferring some centralized administration and land-use rights to local administration and even to individual households.

The common question for these transfers has to do with just how far a government can go with decentralization, and even privatization, of its forests while still satisfying the broader public values that forests provide. One widespread argument against decentralization is that farmers have little experience planting trees and, furthermore, they would not plant trees even if they had the experience. Their time horizons are just too short. China now has more than 20 years of experience against which to test this fundamental perception. A related argument holds that individual farmers will not respond to the natural environmental benefits of reforestation because those benefits are shared widely. This argument anticipates that most of the benefits of reforestation accrue to the farmer's neighbors and even to those who live farther away. Therefore, it concludes that private action is unlikely and that some level of public action will be necessary to provide these environmental services. China has experience against which to judge this argument, as well.

Finally, the lessons from China can be convincing because the evidence is so very clear. China's geographic breadth and tremendous economic growth mean that the data incorporate great variability, and variability in data is a requirement for statistically meaningful results. Imagine, for example, measuring the effects of tree-planting incentives in a region of slow and steady development, perhaps a region like North America's Great Plains, versus measuring the effects of similar incentives in China's grain-producing northern plains. We might observe some success in the Great Plains, but the additional numbers of trees planted would probably be small and it might be difficult to distinguish among a number of market and policy factors that could have induced the additional planting. Now consider an example from China. The number of trees planted in some parts of China's northern plains since 1978 has almost tripled. The market and policy factors that induced the new planting in China may be similar to those affecting the Great Plains but, because China has experienced so much growth, the statistical exercise of separating these factors influencing growth and measuring their effects will probably be more reliable in China. And what is true for this example of tree planting is true for many activities in an economy of great variability, diversity, and growth. As a result, lessons that hold elsewhere in the world are often clearer in a context such as China's.

Background

The rest of this introductory chapter provides a summary of the changes affecting China's forests since 1978, a synopsis of China's forest administration and its geographic organization, and a chronology of China's general reforms and their broad effects. The chapter closes with a review of specific forestry-sector reforms and an introduction to the analytical chapters that follow.

China's Forestry Sector since 1978[1]

Although China's forest history extends over hundreds of years, its current experience is a matter of the past 25 years. In all but the least accessible regions, few trees, let alone forests, survived the destruction of repetitive war and almost continuous resource exploitation since the 1930s. Similar experiences with civil turmoil in other parts of the world cause us to anticipate the consumption of stock resources like standing forests during periods of severe political unrest and economic hardship. China's experience may be an extreme example, if only because its forests suffered for so long—more than 40 years—without much opportunity for recovery.

The central government recognized the problem and addressed it with land-rehabilitation projects in the early 1950s. These initially focused on biological and engineering improvements, including afforestation projects for controlling wind and water erosion and the salinization of soils. Later, the economic adjustments of the early 1960s promoted multipurpose afforestation activities such as intercropping and shelterbelts, which also had the objectives of protecting the environment and improving agricultural productivity.

However, the very limited new forest resources were rapidly depleted during the period of collectivization and industrialization that began in 1958. Collectivization raised doubts about ownership security for forest resources, and thousands of backyard furnaces constructed for steel production consumed most of the small stock of farm trees that remained after the war with Japan, World War II, and the establishment of the communist government. Subsequently, during the Cultural Revolution (1966–1976), large areas of the remaining natural forest were cleared as part of a campaign to expand cropland. Finally, the inventory of state-owned forests was also depleted to support production and birthright employment in state-owned sawmills.

The Cultural Revolution and the political struggles that dominated Chinese society for the decade that followed restricted any real opportunity for economic growth or environmental conservation in the broader economy. Agricultural and forest production stagnated. Per capita annual income in the communes was only 62.8 yuan (about US$25) in 1976; more than one-third of rural households were in debt; and many households lived with insufficient fuel, clothing, and housing. About 100 million people suffered from food shortages (Lu 1986).

At the outset of economic reforms in 1978, China's forests were seriously depleted, and many existing plantations were in their second or third rotations from the same basic stock. Production was declining. Forests covered 82.6 million ha in 1978, but this was only 8.6% of China's total land area or 0.10 ha per capita, among the lowest per capita shares in the world (*China Forestry Yearbook* 1986). Of course, the small per capita stock indicates the likelihood of low volumes of timber harvests and more severe environmental stress than we expect to observe in many developing countries. It also suggests relatively high prices for whatever timber and nontimber forest products were extracted from the remaining forest.

Table 1-1 summarizes conditions in China in 1978 at the outset of the reform period, as well as China's growth through 1998. 1978 marks the introduction of reforms, and it is also a year for which China has a complete inventory of forestland and standing forest volume. 1998 marks China's most recent complete forest survey. Forest surveys consume a lot of resources and each survey requires several years to complete. Therefore, few countries conduct more than one survey per decade. We are fortunate to have benchmark survey dates that coincide so well

Table 1-1. *Economic Growth, 1978–1998*

Growth	1978[a]	1998	Change (%)
GDP (1978 monetary value)			
Aggregate	362	2,312	538
GDP per capita	379	1,869	390
Rural GDP per capita	133	945	610
Forest products (× 1,000 m^3)			
Production			
Logs	51,673	55,557	7
Lumber	11,055	17,876	62
Wood-based panels	1,017	10,563	939
Paper and paperboard	4,390	21,256	385
Imports			
Logs	1,870	4,820	158
Lumber	75	1,679	2,139
Wood-based panels	258	1,977[b]	666
Paper and paperboard	767	5,760	651
Exports			
Logs	28	63[b]	125
Lumber	13	255	1,862
Wood-based panels	10	598	5,980
Paper and paperboard	229	250	9
Forest cover			
Area (million ha)	115	154	35
Standing volume (million m^3)	8,801	10,086	15

[a]Data for imports are from 1981 and data for exports are from 1983.
[b]Data are from 1997.
Sources: China Statistical Yearbook 2000; *China Forestry Development Report* 2000.

with our period of interest.[2] Nevertheless, the 1978–1998 snapshot is imperfect for our purposes because it does not reflect the more recent expansion in log imports that accompanied the Natural Forest Protection Program and a logging ban on the upper watersheds of the Yellow and Yangtze Rivers.

China's real GDP has increased fivefold and its per capita GDP has increased almost fourfold since the introduction of reforms. China's vast rural economy, including forest-based communities and many other households dependent on the forest for some or all of their livelihood, has shown the greatest improvement. Rural per capita GDP has increased a phenomenal sixfold during the same 20-year period (Table 1-1).

Growth in the forestry sector has been more variable and, in some parts, not as impressive. Log production has increased only slightly, whereas lumber production has increased a moderate 60%. The paper industry, on the other hand, has kept pace with double-digit annual growth in the general manufacturing sector, expanding to almost five times its former level. The contrast in growth between lumber and paper is not entirely unexpected. Lumber consumption in most economies is income-inelastic, whereas paper consumption is more income-responsive and may even be elastic over a range of incomes. Therefore, we expect paper production to expand more rapidly than lumber production in order to keep up with the more rapidly expanding demand for paper that generally accompanies economic growth.

The expansion in paper production anticipates a critical handmaiden of early economic development: deteriorating environmental conditions. The pulping technology that is an antecedent to papermaking can be a serious polluter, emitting fibrous solids and chemical waste in its effluents. In China, the pulp and paper industry has become the largest source of rural environmental pollution. Meanwhile, environmental protection has become a serious objective of the central government and, in 1999, President Jiang Zemin announced that any enterprise not in compliance with environmental regulations by 2000 would be closed.[3] More than 2,000 small papermaking enterprises have already been shut down.

The information on nonmarket forest products and services cannot be summarized as easily. Aggregate data on products such as fuelwood, bamboo products, mushrooms, and the many herbs collected from natural forests, and forest-based environmental services such as erosion control, protection of biodiversity, and forest recreation are simply not available. Nevertheless, the evidence that is available is instructive.

Forests are the source of perhaps 40% of rural energy in China, but we are unsure how this estimate has changed over the course of 20 years of development. We are equally uncertain about the effect of China's rapid growth on other nontimber and nonmarket forest products. We can only anticipate that the subsistence-based consumption of many nontimber forest products tends to increase with early development and then eventually decreases as development proceeds and incomes rise further yet. This is a common hypothesis, but its validity has not been demonstrated for China. If it accurately reflects subsistence-based forest consumption in China, then we would want to know the income inflection point at which these consumptive demands on the forest begin to decrease. Has China's economic growth approached or exceeded this point? We don't know.

Although some consumptive demands on the forests may decrease as China continues to grow, we can also anticipate that nonconsumptive demands on the forest will increase. In general, as household incomes increase and overall human welfare improves, recreational uses of natural resources such as forests and demands for all varieties of environmental services also tend to increase. This too is a hypothesis based on general experience, but we are more confident of its validity for China. Casual observations from any of China's mountain or lake resorts reveal a rapid increase in outdoor recreation over the past few years, and the central government's decisions to enforce environmental compliance and establish sustainable development as one of two fundamental strategies for the twenty-first century clearly demonstrate the increasing importance it places on environmental values.[4]

We can conclude that specific demands on the forest have been variable in their relative growth, but the overall importance of China's forests is undoubtedly increasing. How does this expectation compare with aggregate trends in the forest inventory since 1978? Table 1-1 shows that forest cover has increased a modest 35%, while standing forest volume has only increased 15%. The difference is not surprising. New seedlings and new forests grow slowly at first, and their growth in the first few years is generally too small for measurement in most forest surveys. Nevertheless, the expansion in forest cover appears to be proceeding more slowly than the increasing demands for either commercial forest products or nonconsumptive forest-based environmental services.

Furthermore, net imports do not account for the difference between the increasing demand for forest products and the relatively constant forest base. The level of exports has been small compared with domestic production, and the level of imports has not been much larger. Therefore, China's industry has relied on substitute raw materials to absorb the difference; for example, using bamboo as a substitute for wood in some construction activities and agricultural residues as a substitute for wood fiber in paper production. Under most conditions wood and wood fiber are the preferred raw materials, and we can only anticipate that these preferences are another reason why China's forests will become even more important as the country continues to develop.

Table 1-1 summarizes China's rapid growth and the more variable experience of its forestry sector. China's experience has been variable in other respects as well. Reform was experimental, pragmatic, and gradual. There was no overall blueprint. As a result, China has experienced at least three economic cycles during the past 25 years, significant regional variation in the applications of the reforms and in their results, and also differences between broad institutional categories of management. We will summarize these regional differences in forest organization and then trace the chronology of the reforms themselves in the next section.

Forest Organization and Administration[5]

China organizes its official forest statistics by region and administrative responsibility. Table 1-2 records the land area under forest cover for three national forest surveys conducted between 1977 and 1998 in each of six principal geographic regions and according to its classification as either state-owned or collective. The Northeast and the Southwest, followed by the South Central region, contain the largest forests. The map that appears on page xvi of this book outlines the geographic extent of these regions and identifies the provinces within them.

The state-owned forests include approximately 3,000 independent forest farms and about 135 state forest bureaus. The bureaus control their own forest farms, and most of them control their own wood-processing facilities as well. The bureaus were originally set up as logging operations for natural forests, although many of their natural stands are second growth. Many of the independent forest farms were originally set up to manage plantations. The reporting hierarchy for both institutions has changed several times. They currently report to local or pro-

Table 1-2. *Forestland Area*

Region and administration	Forest stands (million ha) 1977–1981	1984–1988	1994–1998	Change (%)
Northeast				
State-owned	21.37	20.55	23.50	9.97
Collective	2.33	3.96	4.20	80.26
Total	23.70	24.51	27.70	16.88
Northwest				
State-owned	2.48	5.30	5.78	133.06
Collective	1.96	2.80	3.19	62.76
Total	7.24	8.10	8.97	23.90
	(4.44)[a]			(102.03)[a]
Southeast				
State-owned	1.32	1.73	2.49	88.64
Collective	10.84	12.01	16.55	52.68
Total	12.16	13.74	19.04	56.58
Southwest				
State-owned	NA	8.15	9.42	
Collective	NA	12.24	17.39	
Total	23.53	20.39	26.81	13.94
South Central				
State-owned	1.78	2.18	2.55	43.26
Collective	10.32	14.92	22.44	117.44
Total	17.35	17.10	24.99	44.03
	(12.10)[a]			(106.53)[a]
North China Plains				
State-owned	13.26	13.11	14.18	6.94
Collective	1.55	2.13	3.43	121.29
Total	14.83	15.23	17.61	18.75
	(14.81)[a]			(18.91)[a]
All China				
State-owned	NA	51.01	62.01	
			(57.93)[b]	
Collective	NA	48.06	67.19	
Total	98.81	99.07	129.20	30.76
			(125.12)[b]	(26.63)[b]

Notes: NA, data not available. China separates the reporting of its forest inventory into "economic forests" (tree crops, such as orchards and rubber), "bamboo forests," and "forest stands" (what most of the rest of the world calls "forests"). The values reported in this table are "forest stands".

[a] In China's second forest inventory (1977–1981), data from some smaller provinces were included in the total but not in the disaggregation into state-owned and collective lands. The first value in the "total" row for the three regions is the value reported in the national forest inventory. The value in parentheses is the sum of recorded state-owned and collective lands for larger provinces in those three regions. The difference is the disaggregated state-owned and collective area in the smaller provinces in the each region.

[b] The value in parentheses is sum of our regional data. The larger value is the official national aggregate.

Source: China Forestry Yearbook 1986.

vincial governments for personnel and planning but both are subject to the forest management guidelines of the State Forest Administration (SFA, the Ministry of Forestry before 1998).

Collective forests are forests that were managed by local forest and agricultural collectives until 1978. Since then, individual households have gradually assumed long-term contractual responsibility for more and more of the collective lands. About 60% of China's total forestland now belongs to the collectives, and individual households manage about 80% of that.

The land areas in state-owned and collective forests have expanded gradually, but the collective forests have grown more rapidly in most regions. Nevertheless, standing forest volume per hectare remains almost three times greater in the state-owned forests, even in the 1998 forest inventory. We would anticipate this ordering because some state-owned forests are inaccessible and biologically mature, whereas all collective forests are accessible and, therefore, almost certain to be harvested at younger ages and smaller volumes as they attain economic maturity.[6] The standing volumes in the state forests of most countries exceed the volumes of their nonstate counterparts for the same reason. However, a threefold difference in volumes per hectare between state and nonstate forests is uncommon. Is this difference in China simply due to the common difficulties of measuring the newest young growth in any forest survey—while the collective forests support a disproportionate share of young stands? Or, is it due to China's accounting system for forests, which includes commercial tree crops such as orchards, which contain less standing timber volume per hectare and are almost totally concentrated on the collective lands? Or, is the difference due to uncertainty and a range of other factors that discourage private investment? We will be interested to observe whether China's collective forest volumes increase rapidly and whether the difference in standing volume between state and nonstate forests diminishes as the effects of the reform incentives for planting and management are more completely absorbed into the measurable growth that will be captured in future forest surveys.

Some discussions of China's forests modify the regional organization of Table 1-2 to reflect management responsibilities more closely. For example, the Northeast and Southwest are sometimes combined (Northeast Southwest State Forest Region, or NSSFR) as the regions of predominantly state-owned forest and as China's major sources of industrial wood. The objective of these state forests has been to support employment in the local mills. Their standing inventory declined because the procurement prices paid by the mills were too low to support reforestation. The Southern Collective Forest Region (SCFR; including the Southeast, South Central, and some of the Southwest from Table 1-2) is the next most productive region. It, too, has suffered from procurement prices that are still too low to encourage reforestation without external financial assistance. The North-Northwest Farm Forest Region (NNFFR) covers the remainder of the country. The western part of the NNFFR is a dry area that supports few forests; the eastern part is a major agricultural area. The only significant forests in this region are government-supported shelterbelts.

Agricultural and other rural policy reforms have affected the SCFR and the NNFFR but they may have had less impact elsewhere. They have led to vigorous agroforestry activities that have significantly altered the landscape in the eastern part of the SCFR. More recent reforms of the SFA and the state forest bureaus plus a few large-scale government plantations may have had greater impacts on the NSSFR.

Chronology of Reform

China's reforms began in the agricultural sector in 1978 and eventually spread to other sectors of the rural and then the urban economy. We can anticipate that the impacts on forestry and the forestry sector occurred largely in response to reforms in other sectors—due to the desire of those in the forestry sector to obtain incentives similar to those that had been successful in other sectors and due to the expanding demands for forest products originating from other, more rapidly growing sectors of China's economy. Therefore, it will be useful to identify the chronology of broader reforms in all sectors as well as of the specific reforms in the forestry sector. It will also be useful to separate the responses to forestry reforms themselves from forestry's responses to reforms in other sectors in the analytical chapters that follow.

The history of reform has been cyclic with initial reforms followed by growth, then inflation and budget deficits in the central government, then retrenchment, and then another round of reforms and growth. China experienced three of these cycles between 1978 and 1994.

1978–1984. Agricultural reforms contained two fundamental elements: the decentralization of property rights from the prior system of centrally administered collectives and production teams, and upward adjustments in the prices for agricultural outputs. Both elements improved the incentives of farm households. The grassroots appeal of both was great, although neither was sanctioned by the central government until September 1980. The original sanction was only for poor, remote, and hilly regions. Nevertheless, the reforms spread rapidly because they appealed to farmers and government officials alike. Both groups found advantage in the diminished transactions costs and risks of government organization that accompanied reforms, as well as in the increased agricultural production that followed rapidly.

The institutional change in property rights began in Fengyang County, Anhui Province, in 1978. It took different forms in different counties. In the most popular version, known as the household responsibility system (HRS), the commune contracted everything (i.e., land, animals, implements, output quotas, and taxes) to its member households. The distribution of land and animals occurred on a largely egalitarian (per capita) basis. The contracts required the households to sell their production quotas at centrally determined procurement prices but permitted them to sell production in excess of these quotas at market prices. This contract system developed slowly until 1981, and then expanded rapidly to 70% of all rural communities by the end of 1982.

The market opportunity stimulated production. Farmers met their quotas with lower quality material and sold their higher quality production in excess of quotas at market prices. Government surpluses accumulated, even as the government raised its procurement prices to levels that were closer to market prices.

Sharp increases in procurement prices followed, and the central government suffered budget and trade deficits in 1979 and 1984. Inflation reached a record 6% in 1980. The aggregate economic conditions caused some rethinking, and the authorities withdrew HRS rights from some households. Nevertheless, these rights extended to almost all agricultural production units by the end of 1984. Compulsory production quotas and the two-price system for outputs were finally abandoned in 1985. Still, all was not perfect. Households saved and invested in durable consumer goods, but agricultural investment lagged, particularly in irrigation. The general suspicion is that this lag was due to uncertainty—uncertainty regarding policy and administration in general, and uncertainty with respect to the durability of the HRS contracts for land in particular. (We will see that these uncertainties became critical issues for forestry as well.) Nevertheless, the overall effect of the initial agricultural reforms was positive and impressive. Land productivity increased 225% between 1978 and 1984, and the productivity of agricultural labor increased by 172%.

1985–1990. A second round of reforms began in 1985. China experienced another cycle of inflation and tighter central controls on expenditures in 1986 before a period of easier credit and expansion beginning in 1987. This round of reforms supported the development of nonagricultural enterprises in rural areas and the extension of the contract responsibility system to the industrial and financial sectors in urban areas. China's leaders first spoke of establishing a market economy during this period.

The central government continued its standing restriction on regional labor mobility, a policy that it expressed as "leave the land but not the countryside." The government provided little direct stimulus for rural nonagricultural development, but it improved the environment for local enterprise by decentralizing the administration of nonagricultural activities and transferring the tax and investment decisions and the administration of environmental controls to the counties. Decentralization encouraged diversification and specialization according to local comparative advantage, as well as regional interdependencies and market relationships that benefited all.

Meanwhile, improvements in agricultural wealth led to an increase in household savings—which were already among the highest in the world. The high level of accumulated household savings provided the financial means for local investment. It financed the new opportunities for local enterprise, known as township and village enterprises (TVEs), that emerged under contract with the local administrative authorities.

The contracts that developed between TVE managers and local authorities were (and still are) similar to those developed under the HRS system for agriculture. Individual enterprise managers contract for fixed responsibilities, and their remuneration is tied to performance. Profits from performance greater than contracted quotas is usually shared with the local government. Therefore, the local governments have their own incentives to ensure the conditions for the growth of these enterprises. The incentives are limited, however. Local governments do not have the financial resources of the central government. Therefore, they insist on tighter budget constraints. Within the administrative discretion of the local authorities, the TVEs became almost entirely market-oriented by the mid-1980s.

The rural nonagricultural sector expanded simultaneously with agriculture throughout the early 1980s. Then, as the rate of agricultural growth began to decline, the TVEs entered their period of most rapid growth. Growth in China in general slowed in 1986 as some central authority was reimposed following another period of inflation. However, for the full period between 1978 and 1998, the rural nonagricultural sector grew at a rate of more than 20% annually. The TVEs became an outlet for underemployed agricultural labor, absorbing new labor at an annual rate of 15%. Their production surged, and their share of gross rural product increased from 26% in 1984 to 45% in 1992. The TVEs accounted for one-half of the increase in China's per capita income during the 1980s.

Two industries of special interest to forestry, wood products and pulp and paper, are generally located in rural areas. The firms in each are small. Yet TVEs accounted for 90% of the enterprises in the paper industry by the late 1980s, and production grew more rapidly in the TVEs than in the generally larger state-owned enterprises (SOEs) for both industries. TVE production more than doubled between 1978 and 1989 for both industries.

New problems emerged along with the success of the TVEs. Some expanded more rapidly than their markets justified. Some local governments tried to protect their own shares of revenue flows from these enterprises by introducing barriers to interregional trade. Excess capacity resulted and this led to the excess use of some productive inputs, including the excess conversion of agricultural land and forestland. And, like most other examples of early industrial growth, air and water quality suffered as well.

The first signs of change in the previously egalitarian distribution of income appeared during this period. Generally, agricultural income is less variable than the income from some nonagricultural activities, including income from the wood products and paper industries. Therefore, growth in the TVEs increased the income disparity within regions. Furthermore, the success of the TVEs was distributed unevenly across regions and this created regional income disparities. Regions with better infrastructure, more capital, better education, and better access to major urban areas and outlets for foreign trade grew faster. As a result, household incomes and local government revenues and budgets for social services became more uneven. Nearly all households and counties benefited, but they did not benefit uniformly.

Finally, the industrial- and financial-sector reforms of this second period also emphasized the transfer of the contract responsibility system that had been so successful in agriculture. In this case, the transfer was to the largely urban state-owned industrial and financial sectors. The central government allowed the managers of SOEs discretion on variable inputs and on output levels and mixes. The government decreased its budgetary role in most SOEs but continued to control most capital investment decisions. Employment in these firms remained a birthright: children of employees were entitled to work in the same SOE that employed their parents. Despite these constraints on decisions about labor and fixed capital, managers improved input efficiency sufficiently to increase productivity at an annual rate of more than 15% for the full manufacturing sector.

1991–2000. A third period of reforms followed the events of Tiananmen Square in mid-1989, a crackdown by hard-line political leaders, and then a cooling-off period that extended until 1991. The reforms of this period continued the previous period's focus on SOEs, and the role of the central government continued to decline.

Beginning in 1991, the government allowed the sale of some SOEs and permitted others to go out of business. Seventy percent of small SOEs were privatized by 1997. The government also allowed managers to release redundant employees beginning in 1995 and, by 1998, one-fifth of all employees of SOEs (7 million people) had been released. Most either found employment in TVEs or lived off their severance allowances. Meanwhile, the government itself downsized, reducing its budgetary role in most SOEs, simplifying the tax system, and decreasing the number of civil servants from 8 million to 4 million. Most retail prices became market-determined. Managerial discretion and productivity continued to increase as a result. (The industrial sector grew at an annual rate of 6% in the early 1990s but declined slightly in the late 1990s as redundant labor was released and first East Asia, then the world, experienced economic decline [*China Statistical Yearbook* 2000].)

Finally, the government liberalized the markets for tradable goods. It allowed the retention of earnings from foreign trade in 1984, removed direct export subsidies in 1991, and eliminated centrally regulated exchange rates in 1994. It officially allowed joint ventures with foreign capital as early as 1979, but this opportunity did not become important until the late 1980s and early 1990s when the government approved the location of joint ventures in some coastal counties.

Household incomes continued to increase, and so did savings. Accumulated savings were 6% of GDP in 1978 and a phenomenal 62% of GDP in 1998. The reservoir of savings may reflect current investment opportunity, or it may reflect household decisions to protect against a future that many still perceive as uncertain. In either case, it indicates a deep well of opportunity for future investment and growth. The first 15-year period for most HRS contracts expired in the mid- to late 1990s. The government did not alter its policies when these contracts expired. Therefore, agricultural households may now have greater confidence in the stability of government policy. Is this the confidence they need to begin investing? Time will tell.

Despite these reforms, the central government remains a major actor in the economy. Most of the largest firms are still state-owned, and many unprofitable SOEs remain in operation. The government's role is an active one in other respects as well. It continues to restrict labor mobility between regions and to the cities. It aggressively seeks international acceptance of its trade policies. And it has recently begun taking a stronger view toward environmental protection—charging fees for emissions, discussing tradable permits, and even closing some firms that are not in environmental compliance, including several thousand small paper mills.

Forestry-Sector Reforms and Chapter Overview

China introduced a number of reforms that were specific to forestry. This section introduces the most important, reflects on their place within the context of China's overall economic reforms, and considers the questions they raise for the subsequent chapters of this book and for the design of forest policy around the world. (The Appendix Table provides a chronology of important central government forest policy decisions.)

The Three North Forest Protection Project (also known as the Green Great Wall) was the first large government program initiated after reforms began. The Green Great Wall is a shelterbelt reforestation program that was designed to combat desertification and wind erosion. Its construction began in 1978 and continued throughout the reform period, eventually extending to 4 million km^2. This large program became a component of larger government policy to accelerate afforestation in 1980.

The massive government reforestation program was impressive, but successful agricultural policy clearly provided the earliest incentives for forest policy reform and development. The agricultural reforms that began in 1978 produced a phenomenal increase in agricultural production. Two effects on forestry followed directly. First, some agricultural households, once they received HRS rights to their lands, also invested in trees. This effect was most noticeable in the grain-producing North Central Plains where wind erosion was a serious problem that could be modified by planting trees and where forestry had been only a minor crop that was largely overlooked by the government's regulatory authorities.

Second, forest-dependent households observed agriculture's success with HRS rights, and the members of forest collectives demanded similar rights for themselves. A comparable system known as the contract responsibility system (CRS) was introduced in 1981 for forest cooperatives and forest households. This system became the third component of the "Three-Fix" policy for forestry (stabilizing the rights to and ownership of forests and mountains, identifying boundaries of household plots, and establishing a forest production responsibility system) that was implemented in the SCFR between 1981 and 1985.

The share of households participating in the CRS rapidly increased to 55% of those in forest collectives by 1983, and by 1984 approximately 30 million ha had been designated for private management by about 57 million households. The reforms continued when the government liberalized markets in the collective forest areas in 1985 by abolishing the state monopoly on the purchase, sale, and distribution of logs. Real timber prices increased 43% in the SCFR that year.

Differences existed, however, even between counties, in the pace and extent of these forestry reforms. Many local forest authorities doubted the willingness of individual households to make the long-term commitment necessary to grow trees. With the inflation of 1986 and the general retrenchment that followed, some local authorities withdrew CRS rights from some households until only 52% of households in forest collectives possessed these rights in 1988. The state monopoly on timber purchase was also reestablished in the SCFR in 1987. (The state monopoly was repudiated and an open market policy was reconfirmed once more in 1993.)

The next two chapters of this book examine these experiences and other characteristics of the devolution of authority and responsibility from the central government to local authorities and local households. Liu and Edmunds, in Chapter 2, describe the change in the institutional arrangements for property rights to the former collective forestlands and the shifts in oversight responsibility for these lands from central to regional and local governments. They rely on national forest survey data and their own research experience in 22 villages in China's South and Southwest to illustrate the impacts of these changes. (Seventy percent of the affected forests are located in the South and Southwest.) Liu and Edmunds show that the effects of reforms have changed over time; for example, from an initial period of rapid deforestation in the early 1980s to gradual reforestation and forest growth in the 1990s. They also describe some of the variation in impacts between households and between regions and they suggest some of the remaining opportunities for policy reform in the sector.

Despite the devolution of authority and the growth in both forestland and forest stock, investments in forest management remain more constrained than those in comparable agricultural activities. Individual households now possess the rights to the products of their household plots, and most of the controls on the inputs of productive activities were removed by the late 1980s, but the local forest bureaus maintain the right to limit harvest levels and shipments of forest products.

Liu and Edmunds point out that various levels of government also impose an array of timber taxes and charges that account for more than 50% of the market value of timber in some provinces. The relevant agencies began increasing these fees shortly after timber prices began to rise in response to market liberalization in 1985. The agencies saw an opportunity to increase their own revenues, and they took advantage of it. Since then, taxes and charges have risen more rapidly than timber prices. This contention is well-known among Chinese foresters and also in the English literature (e.g., Bruce et al. 1995; Albers et al. 1998; Yin 1998; Li 1999; Zhang 1999; Chen et al. 1999, 2000; Zhang 2001). Liu and Landell-Mills remove it from the realm of contention in Chapter 3. They provide a detailed description and numerous local examples of these disincentives for one region, the SCFR. Their example is particularly appropriate because the SCFR is a major region for commercial forestry and, along with the eastern part of the NNFFR, one of two large regions where individual households are the most important forest managers.

The second set of two chapters continues the inquiry into forest investment, but with a more analytical approach. In Chapter 4 Yin conducts econometric assessments of three questions that arise in the discussions of Chapters 2 and 3. In Chapter 5, D. Zhang extends the inquiry more broadly with a review of the evidence on all sources of new forest investment since 1978, including state and external sources, as well as the household sources that were the focus of Chapters 2 and 3.

Yin's chapter begins with an examination of two central features of reform described in Chapter 2. Yin estimates the effect of CRS rights on household timber production and investment and then continues with an estimate of the effects of the uncertain policy environment on the inventory of forests and forestlands. Surely, improved land-use rights must have led to an increase in forest investment. Just as surely, the act of giving CRS rights for forestlands and then withdrawing them from some households must have created an uncertain environment for investment, and that uncertainty must have affected forest management. Yin demonstrates these effects with contrasting observations from adjacent parts of the country. In the northern plains, forestry activities were less regulated, and households responded rapidly to new incentives to plant trees. Forestry has traditionally played a larger role in the economy of the provinces just south of this region. However, the same government policies were applied in a less even and predictable manner in the southern region. As a result, households in this second region modified their behavior with regard to both planting and timber harvesting to the detriment of longer-term forest management.

These observations on improved private land-use rights are important in light of the global attention given to potential transfers in forest tenures, especially transfers from state lands to local communities. The authors of Chapter 2 remind us, however, that any sort of major modification of land-use rights is likely to be complex and subject to much experimentation and modification before successful local arrangements develop. As in Chapter 2, it is implicit in Yin's results that individual private rights are superior to community rights, at least in some cases. Both chapters' observations on policy uncertainty introduce an issue that most discussions of forest policy overlook but one which is critical in the many countries of the world that are subject to rapid economic and political change. That is, although the flexibility to respond to local differences is desirable, frequent policy experimentation creates uncertainty and imposes significant environmental and social costs.

The authors of Chapters 2 and 3 identify the serious disincentive of China's burdensome taxes on timber. This raises Yin's third question. How can the expansion in forest cover on the collective lands be explained in the presence of confiscatory taxes and regulations? For at least one region, Yin demonstrates that even poor farmers with new HRS rights but short time horizons comprehend the advantage of improving the agricultural environment. Farmers in the northern plains planted trees and obtained almost immediate benefits in the form of erosion control and improved soil conditions—reflected in measurable increases in agricultural productivity. This evidence urges caution for the forest authorities in China and elsewhere who doubt that farmers will plant trees of their own accord and also for those who presume that public action is necessary to obtain environmental benefits. Farmers do appreciate the benefits and they do plant trees—at least in some cases.

D. Zhang takes a different approach to investment in Chapter 5. He examines four sources of forest investment: farm households, the government, private nonfarm investors, and foreign investors. He examines the incentives and tracks the performance of each. His observations for farm households are consistent with those of the prior chapters. Zhang's more interesting contributions lie in the observed patterns of government and private nonfarm investment. Government funding has tended to shift toward large silvicultural investments over the last 10 years. However, it has not produced much response in terms of growth in standing forest volume. Its results have been unlike those for farm household investment. Private nonfarm investment, a third category, is small but growing. Zhang traces the returns on investment for private firms and determines that their investments are riskier than those in other manufacturing-sector activities. Apparently, uncertain policy, high taxes, and insecure property rights deter further growth in these investments—much like they deter further growth in farm household investment.

Zhang also speculates on the future. In the presence of general economic growth and the absence of significant increases in forest production, the wood-consuming construction and paper industries must obtain their raw materials elsewhere, probably from imports and from substitute materials. Zhang speculates on these alternative sources and the adjustments they would impose on China's economy.

The chapters to this point focus on reforms that were designed for their prospective impacts on the agricultural or forestry sectors. The third set of two chapters examines the contention that market and policy activities outside the forestry sector are as important to forestry and the markets for forest products as forest policies themselves. In Chapter 6, Rozelle, Huang, and Benziger first discuss the trends in China's regional forest inventories between 1978 and 1994 and then examine those trends as econometric functions of forest and agricultural markets and policies and as a function of China's general economic performance. They add a discussion of China's forest data, the basic data on which their chapter and several others rely.

In Chapter 7, Y. Zhang, Uusivuori, Kuuluvainen, and Kant conduct an assessment of changing forest conditions on the island province of Hainan. Their assessment is comparable to the assessment in Chapter 6 in that it is an econometric search for the broad market and policy determinants of forest cover. However, Zhang et al. obtain more instructive results by separating effects on managed stands from effects on natural stands. Their analysis reflects a narrower geographic range but a better identified single market than that of Chapter 6, and its observations are more robust as a result.

The authors of Chapters 6 and 7 conclude that market and policy activity in the broader economy are important determinants of forest cover and standing forest stocks. They also emphasize the importance of distinguishing between natural forests and managed forests. These two categories of forest provide different forest resource services, and the markets and policies that affect one category can have different and even opposite effects on the other. Increasing prices, for example, encourage increased harvesting and reduction in the natural forest area, but they are an incentive to expand managed forests.

Both components of this observation are important lessons for all of forestry, globally as well as in China. Surely, it is important to specify first which forest resource services a policymaker intends to enhance, then to identify the categories of forests that support them and, finally, to design policy to target only the selected forests and their resource services. Just as surely, spillover effects from markets and policies outside the forestry sector itself often affect forestry, with different impacts on different categories of forests. Errors in judgement and policy are common when analysts and policymakers disregard these fundamental observations.

The final analytical chapters introduce two emerging concerns: income distribution and the environment. There is no well-developed literature on forestry's role in these issues in China—perhaps because they are relatively new to the policy discourse. These chapters offer a brief introduction to some of their important elements.

Most household incomes in China have risen substantially since 1978, but the disparities between households also have increased, within provinces and even more greatly across broad regions.[7] Many rural households remain poor. Furthermore, among 592 counties officially designated as poor, 496 are in mountainous areas where various forestry activities are major sources of household welfare. The question is whether forest policy reforms and forest development can alter these conditions. Have they helped the poorest households and counties and mitigated the increasing income disparity?

In Chapter 8, Ruiz Pérez, Belcher, Fu, and Yang examine one forest product, bamboo, and the effect of its growth on income classes and general regional development. Bamboo, in all its many uses, is probably the third-most-important forest product in China (after fuelwood and commercial timber). It is a particularly useful product for the inquiry of this chapter because bamboo has not been subject to as many external policy restrictions as products like commercial timber. Therefore, its recent history is more representative of the impacts that unfettered forestry-sector development could have on household incomes and general regional development.

Ruiz Pérez and his coauthors challenge us to rethink the conventional wisdom. Who benefits from bamboo (or forest) development? If many benefit, then which income groups benefit the most? Can we reasonably expect forestry-sector reform to benefit the poorest? Is additional forest-based income the benefit itself, or does the diversification that this income adds comprise the greater benefit for local households? And regarding regional economic development, just what are the limits forestry as a leading sector for regional growth? This chapter does not set out to be comprehensive. The data are extensive, but only for one product and for one region. Nevertheless, the detail in the analysis should cause us to consider carefully each of these questions for other forestry activities in other regions of China and the world.

The changing forest environment is the final issue addressed in this book. In general, poor countries draw on natural resources as capital that can be consumed to promote economic development. Unfortunately, resource depletion is one outcome. Fortunately, as they develop, some countries become more aware of and more concerned with resource depletion and other environmental problems, and act to mitigate them. Furthermore, development itself may create the financial means to address these problems.

China's environmental problems are serious, but China has begun to address some of them. The list of policies and policy objectives in appendix Table 1-A reflects the increasing attention that the central government has directed toward environmental issues in forest management since the late 1980s. Environmental management has gradually become a central responsibility for all forests, but particularly for the state forests.

Sayer and Sun, two experienced environmental analysts, discuss the topic in Chapter 9. The data for environmental assessments of China's forests are sparse at best. Therefore, Sayer and Sun survey general impressions before focusing on one representative topic, the loss of biodiversity that necessarily occurs as China depletes its natural forests. China's large expanse of natural forest may support the largest range of biodiversity of any country in the world. However, the depletion of its natural forests suggests a large loss of biodiversity as well. Sayer and Sun reemphasize the point from Chapters 6 and 7 about distinguishing between managed and natural forests. Managed forests support some environmental services and they play a crucial role in controlling erosion, but they are not as important as natural forests for the protection of critical habitats and general biodiversity.

The central government has identified the environment as a focal policy issue for the twenty-first century, and it has begun enforcing compliance with environmental regulations on polluting industries. As it begins to add the forest environment to its policy concerns, Sayer and Sun emphasize the importance of the physical sources of each forest-based environmental problem. It is insufficient to presume that comprehensive rules such as increasing forest cover in general, or even saving all natural forests, will solve all forestry problems. This instruction is entirely consistent with the observations of Rozelle et al. and Y. Zhang et al. in Chapters 6 and 7 that managed and natural forests provide different resource services and respond differently to market and policy incentives. Surely, some natural forest habitats are more important than others. Surely, plantations can address some environmental problems. Surely, the problems of forest environments are as real and as far-reaching as Sayer and Sun suggest in Chapter 9. And surely, these problems and their resolutions are as complex as the forest environments themselves. Better data and more thorough analyses are needed to assess the relative magnitudes of these problems and the most effective means of addressing them.

The final chapter reviews the findings of the empirical chapters, summarizes their observations for China, and explains these observations within the context of global forest policy. The institutions that define land tenure and the means of delivering tenurial rights is one great theme of China's experience—and this book. The impacts of spillovers from policy reform in other sectors and from general economic growth is another.

Chapter 10 also reflects on those new issues that arose subsequent to the analyses of the other chapters and those additional issues that are becoming important only now, in the twenty-first century. We have identified the importance of income distribution and the environment, and one must carefully think through the role of forestry in these issues.

Forest environmental issues are particularly complex. For example, the pulp and paper industry is the source of 10% of China's industrial wastewater emissions and one-fourth of its chemical oxygen demand. This industry currently uses agricultural residues at a three-to-one ratio over wood fiber. Yet the papermaking process is less damaging to the environment when it uses wood. Would it be better to increase the proportion of wood fiber—thereby further depleting China's forests? Does it matter that greater use of wood fiber would require increasing wood imports—thus increasing the resource cost in other countries? In another example, China's new Natural Forest Protection Program (NFPP) and its logging ban on the upper extents of the Yellow and Yangtze River watersheds is intended to restore the upland environments and limit erosion and downstream flooding. Are these expectations reasonable? The NFPP includes provisions to assist the financially strapped forest authorities as their responsibilities shift to the ecological functions of the forest, and also the forest workers who will be displaced. What are the reasonable expectations for these activities? How much assistance will be necessary and how successful will it be?

Finally, tenure continues to be an issue, sometimes with a distributive theme, sometimes with an efficiency theme. Land-use rights have been reallocated yet again in some of the regions used as examples in this book. Rights were reallocated on local demand and to maintain a degree of household equity. We can anticipate the effects, but they have been largely ignored in the literature outside of China. Meanwhile, restrictions on household harvests and shipments as well as high levels of forest taxation continue to constrain household forestry incentives. The central government recognizes these issues, but local administrators will largely determine whether future growth in China's forestry sector is characterized by imports and product substitution or by domestic production.

Appendix Table. *Policy Reform in China's Forestry*

By Jiurong Zhang and Xuhe Chen

Date	*Title*	*Issued by*	*Principal objectives/impacts*
3/1980	Introduction of Vigorous Development of Afforestation	Central Committee of CPC, State Council	Accelerates the greening of China, promotes national afforestation and greening.
3/1981	Decision on Several Issues Related to Forest Protection and Development (the "Three-Fix" policy)	Central Committee of CPC, State Council	Sets rules and policies for forest protection and development. For forestry reform: determines the ownership of mountains and forests, designates mountain slopes for household use, and defines a forestry responsibility system. Ownership certificates were issued for 97 million ha of mountain forests (of which 3 million ha were designated as privately managed mountains) for about 57 million rural households.
10/1984	Decision on Reform of the Economic System	The 3rd Plenary Session of the 12th Central Committee of CPC	Shifts the priority of reforms from rural to urban and the emphasis to state-owned enterprises, market development, and price reform.
1/1985	The Forest Law of the People's Republic of China (China's first forest law)	People's Congress	Formally recognizes the division of forests between the state and collectives. States general objectives for forest management, establishes timber harvest quotas, and requires shipping permits.
1/1985	Ten Reforms To Further Stimulate the Rural Economy	Central Committee of CPC and State Council	Abolishes the state monopoly (the Unified Procurement System) for timber purchase, sale, and distribution in collective forest regions, and opens the markets for wood.
5/1986	Implementation Rules for the Forest Law of China	State Council	Implements the forest law; protects forest resources and develops the forest industry.
1987	Enhancement of Forest Resource Management and Interdiction of Ruinous Cutting in the Southern Collective Forest Region	Central Committee of CPC and State Council	Halts ruinous cutting; implements the responsibility system of government leadership at all levels in the protection and development of forest resources. Resumes the state monopoly on timber purchased in the Southern Collective Forest Region.

1988	The Provisional Regulation of State-Owned Enterprise Management Responsible Contracting System	State Council	Improves control over the consumption of forest resources and improves the financial operation of state-owned forests. These enterprises were requested to adhere to annual timber cutting quotas; promote reforestation; diversify management and the comprehensive use of resources; and enhance production safety, forest protection, and fire prevention.
5/1989	Notice of Strengthening the Management of Logging with Certificates	Forestry Ministry	Reinforces the logging quota system and the requirement that forest growth had to exceed timber harvest levels.
1993	Decision on Several Issues Concerning the Establishment of the Socialist Market Economic System	The 3rd Plenary Session of the 14th Central Committee of CPC	Accelerates introduction of the market system and reform of taxation, finance, prices, and rules for foreign investment.
1995	China's Agenda for Forestry Action Plan in the 21st Century (first environmental objectives for forest management)	Forestry Ministry	Presents, in a general way, the development plan for different elements of China's forest sector.
1/1995	Notice of Implementing the System of Using Forest Land with Certificates	Forestry Ministry	Requires certificates for all forestlands. Differentiates forestland from other land uses. Effectively restricts conversion to other land use and ensures that this land will be regulated in accordance with forest law.
8/1995	General Outline for Restructuring the Forestry Economic System	Forestry Ministry	Pushes forest enterprises toward the market, mobilizes and enhances financial support to forestry through tax policy, strengthens infrastructure development, reduces the role of government, and reinforces administrative support for forestry.
10/1995	Instructions on the Decision about Accelerating Progress by the Central Committee of CPC and State Council	Forestry Ministry	Establishes guidelines of reform for science and technology education, set forth science and technology as the guiding ideology for promoting forest development and set the major mission for the near future through 2010.

continued on next page

Appendix Table. *Policy Reform in China's Forestry (continued)*

Date	*Title*	*Issued by*	*Principal objectives/impacts*
5/1996	Experiment on the Reform of the Development of Classified Management of Forestry	Forestry Ministry	Promotes the reform of classified management and operation. Classifies forests according to commercial, environmental (public interest), and mixed objectives.
9/1996	Decision on Several Issues Related to Deepening the Reform in the State-Owned Forest Farms	Forestry Ministry	Emphasizes two points: (1) classified management reform must differentiate between commercial and environmental forests and should be appropriate for local economic and social development conditions and (2) industry structure of forestry farm must be optimized.
1998	National Ecological Environment Establishment Program	State Council	Continues the shift toward environmental justifications for state forest management.
4/1998	Amended "Forestry Law of the People's Republic of China"	People's Congress	Legalizes transfer rights for family plots and extends the period of user rights, essentially stabilizing forest tenures in collective forests; emphasizes that the principal role of forestry is to provide environmental services.
8/1998	Notice on "Protecting Forest Resources, Interdicting Ruinous Deforestation, and Occupation of Forest Land"	State Council	Protects forest resources by preventing the conversion of forestland for agriculture and other development purposes.
8/1998	National Forest Protection Plan	State Council	Bans logging in natural forests at the upper reaches of Yangtze River and upper and middle reaches of Yellow River, reduces timber production of state-owned forest farms in the Northeast and inner Mongolia by 19.91 million m^3, establishes 12.7 million ha of plantations, and redirects and resettles 740,000 excess workers.

Notes: CPC, Communist Party of China; ha, hectares.

Notes

1. Many have reviewed and assessed this experience. We cross-checked the material in this section with reviews in Ho 1994, Chai 1997, Linge 1997, Chen 1999, Du 1999, Qian 2000, Zhang 2001, Zhang and Chen 2001, and Wang and van Kooten 2001.

2. China conducted its first national forest survey between 1972 and 1976, its second between 1977 and 1981, its third between 1984 and 1988, its fourth between 1989 and 1993, and its fifth between 1994 and 1998. As a matter of nomenclature, forest surveys collect the information. The assembled information is called a "forest inventory". The terms "survey" and "inventory" are often interchanged.

3. In his address to the annual workshop on Population, Resource, and Environment on March 13.

4. Premier Zhu Rongji, in the Government Report to the National People's Congress on March 5, 1999.

5. Richardson 1990 is the standard English language reference for the natural science, geographic variation, organization, administration, and production practices of China's forestry prior to the late 1980s.

6. In fact, the inventories of *accessible* state-owned forests were also drawn down severely. Of 131 state-owned forest bureaus in the Northeast and Southwest in 1978, 25 had exhausted their mature wood resources by 1993.

7. China's Gini coefficient, a measure of its household income inequality, rose sharply from approximately 0.20 in the late 1970s to 0.30–0.35 in 1990 (Chai and Chai 1994, Chen et al. 1995). It may be around 0.40 today. Demographers often consider a coefficient that is much greater than 0.40 as indicative of serious maldistribution and potential social unrest.

References

Albers, Heidi J., Scott D. Rozelle, and Li Guo. 1998. China's Forests under Economic Reform: Timber Supplies, Environmental Protection, and Rural Resource Access. *Contemporary Economic Policy* 16: 22–33.

Bruce, J.W., S. Rudrappa, and L. Zongmin. 1995. Experimenting with Approaches to Common Property Forestry in China. *Unasylva* 46(180): 44–48.

Chai, Joseph. 1997. *China: Transition to a Market Economy.* Oxford, U.K.: Oxford University Press.

Chai, Joseph C.H., and Karin B. Chai. 1994. Economic Reforms and Inequality in China. *Revista Internationale di Scienze Economiche e Commerciali* 41(8): 675–96.

Chen, Shao hua, Gaurav Datt, and Martin Ravallion. 1995. *Is Poverty Increasing in the Developing World?* (Data appendix, updated version). Policy Research Department, The World Bank. 40: 359–376.

Chen, Xie, Guoqiang Feng, and Xianbo Wu. 1999. Investigation and Study on Forestry Taxation and Charge-Based on Hunan Province. *Forestry Economics* (China) 6: 26–35.

———. 2000. Investigation and Study on Forestry Taxation and Charge-Based on Hunan Province. *Forestry Economics* (China) 1: 21–28.

Chen, Y. 1999. Decentralization in China: Background, Current Situation, and Its Impact on Economic Development. In *Social Sciences: Chinese Academy of Social Sciences Forum*. Beijing: China Social Sciences Publishing House, 47–58.

China Forestry Development Report. 2000. [In Chinese] Beijing: China Forestry Press.

China Forestry Yearbook. 1986. [In Chinese]. Beijing: China Forestry Press.

China Statistical Yearbook. 2000. [In Chinese]. Beijing: China Statistical Press.

Du, Y. 1999. Transition of the State-Owned Enterprise System. In *Social Sciences: Chinese Academy of Social Sciences Forum*. Beijing: China Social Sciences Publishing House, 38–46.

Ho, S. 1994. *Rural China in Transition*. Oxford, U.K.: Oxford University Press.

Li, Mingfeng. 1999. Problems and Reform Recommendations of the Forestry Taxation and Charges in Guangxi Autonomous Region. *Forestry Economics* (China) 1: 56–63.

Linge, G. (ed.). 1997. *China's New Spatial Economy: Heading Toward 2000*. Oxford, U.K.: Oxford University Press.

Lu, X.Y. 1986. *Research on Responsibility Systems*. Shanghai, China: People's Press of China.

Qian, Y. 2000. The Institutional Foundations of China's Market Transition. In *Annual World Bank Conference on Development Economics 1999*. Washington, DC: World Bank, 377–398.

Richardson, S.D. 1990. *Forests and Forestry in China*. Washington, DC: Island Press.

Wang, S., and J.C. van Kooten. 2001. Forest Policy in Post-1978 China: A Mosaic in Transformation. Paper presented at the Symposium on Policies Affecting China's Forestry Sector, Dujiangmin, June 20–23, 2001.

Yin, Runsheng. 1998. Forestry and the Environment in China: The Current Situation and Strategic Choices. *World Development* 26(12): 2153–2167.

Zhang, J., and Xuhe Chen. 2001. Review and Prospects of China's Forestry Reform. Paper presented at the Symposium on Policies Affecting China's Forestry Sector, Dujiangmin, June 20–23.

Zhang, Xiaojing. 1999. The Forestry Taxation and Charges Issue in the Collective Forestry Area of Fujian and Jiangxi Provinces. *Forestry Economics* (China) 6: 36–45.

Zhang, Yaoqi. 2001. Economic Transition and Forestry Development: The Case of China. Ph.D. thesis. Helsinki, Finland: University of Helsinki, Department of Forest Economics.

CHAPTER 2

Devolution as a Means of Expanding Local Forest Management in South China

Lessons from the Past 20 Years

Dachang Liu and David Edmunds

CENTRAL GOVERNMENTS have experimented with mechanisms to transfer responsibilities and rights for forest management to local governments, private industries, community-based organizations, and individual households at least since the 1970s (e.g., Rondinelli 1981, Rondinelli et al. 1989, Manor 1999, Agarwall and Ribot 2000). Perhaps no country has made more dramatic steps toward devolving the authority for forest management than the People's Republic of China.

China initiated a series of policy reforms in the early 1980s that affected the administration, marketing, and investment infrastructure of the forestry sector as well as the tenure of its collective forests. These reforms were implemented across the entire country. They authorized the decentralization of authority from central to local governments, thereby enabling counties, townships, and administrative villages to make an increasing number of decisions independent of directives from above. The reforms also transferred some forest resources to individual households and empowered households to make more management decisions for themselves.

Many have examined this experience (e.g., Menzies and Peluso 1991, Sun 1992, Bruce et al. 1995, Albers et al. 1998, Yin and Newman 1997, Song et al. 1997, Zhang et al. 2000). Our objective in this chapter is to draw on the key lessons from their assessments and add the experience of our own research from 15 villages in the provinces of Hunan, Guizhou, and Yunnan and our research on forest degradation and rehabilitation from seven villages in Guangxi, Hunan, Yunnan, and Zhejiang (Liu et al. 2002). Our geographic focus is China's South and Southwest, two regions that account for 70% of China's collective forestlands (*China Forestry Yearbook* 1990). We will draw on data from official forest surveys, information from interviews in our villages, and secondary data from local forestry annals.

We intend to review the institutional components of devolution as well as the changes in forest cover and stand composition that followed the reforms, but our analytical focus is on lingering problems: the flows of timber and environmental services from the forest, the flow of long-term investment into forestry, and rising social inequality in the countryside. Our explanations center on resource tenure, regulatory, and marketing institutions; the local capacity to engage in forest management; and government roles and policy implementation at the village level.

The chapter contains four sections. The first summarizes the institutional context for the reforms themselves. The second examines the effects of the devolution of authority and responsibility in terms of forest cover, stand composition, and impacts on human welfare. The third seeks explanations for the effectiveness or ineffectiveness of the new policies, and a final section presents our conclusions and policy recommendations.

Devolution of Collective Forest Management

Prior to 1981, China's forests were administered in two broad categories: state-owned forests and collective forests. They remain officially classified within these same two categories today, although forests in the latter classification have been separated into several categories of nonstate management. Nationally, the collective forests account for 61% of total land area under forest cover. The proportion is higher in the Southwest (Yunnan and Sichuan) and South (Anhui, Fujian, Guangdong, Guangxi, Guizhou, Hainan, Hubei, Hunan, Jiangxi, and Zhejiang). Collective forests account for 64% of all forests in the heavily forested Southwest and 92% of all forests in the South (*China Forestry Yearbook* 1990). The larger shares of collective forest in these regions have implications for the local people who depend on them as well as for forest management itself.

China began its experiments with new management forms for its collective forests in 1981. The experiments followed the pattern of successful agricultural reforms in their focus on new arrangements for local land and resource tenure. Three categories of management and tenure were established: family plots and responsibility hills (both managed by individual households) and collective management, which continued in modified forms (Central Committee 1980, 1981). Table 2-1 summarizes the characteristic tenurial arrangements, land types, geographic occurrence, and trends in aggregate land area over time for each category. By 1987, 70% of the collective forests in seven southern and southwestern provinces (Anhui, Fujian, Guizhou, Hubei, Hunan, Yunnan, and Zhejiang), plus the northern province of Henan, were managed as either family plots or responsibility hills (*China Forestry Yearbook* 1989).

Family Plots

The collectives distributed denuded forestland and brush to households as family plots (*ziliushan*) on the condition that the households would plant trees. Ownership of the land remained with the collective, but the households enjoyed land-

Table 2-1. *Postdevolution Tenurial and Management Arrangements for Nonstate Forests*

	Tenurial arrangement	*Types of land affected*	*Primary decisionmakers (for production and management)*	*Most common occurrence*	*Trend in area covered*
Family plots (*ziliushan*)	The collective owns the land, but its use rights are distributed to households. Trees planted on the land are property of the households.	Denuded forestland and brush	Households (what and where to plant, when and where to harvest and sell nontimber forest products)	Throughout the country	Expanding from 1980 to 1987, steady since
Responsibility hills (*zerenshan*)	The collective owns the land and trees.	Existing forests and lands on which forests grow	Shared between the collective and households	Throughout the country	Decreasing from 1984 to 1990, steady since
Modified collective management (including collective forest farms)	Collective ownership.	Existing forests and lands on which forests grow	Village leadership, but with greater input and participation from villagers than before reforms	Most areas of China	Expanding slightly from 1987 to 1990, steady since
Shareholding system (government dominant)	The collective owns land. Tree tenure has been unclear, although the name implies household ownership.	Existing forests and lands on which forests grow	Board of trustees made up of village leaders and representatives of other shareholders	Fujian, Guizhou	Steady

use rights for an unspecified period of time and owned the trees they planted. Households were allowed to select the species for planting, to make a number of management decisions, and to harvest nontimber products.

Transfers of the use rights to these lands were not allowed. Nevertheless, households began transferring their rights in the form of leases in the early 1990s, and the Revised Forest Law of China legalized these transactions in 1998. Timber management remained strictly regulated even after 1998, with state permits required before the cutting and shipping of timber and heavy taxation of timber revenues.

Nationwide, more than 31.3 million ha or more than 1% of China's forestland was distributed to 57 million households by the end of 1984, when this component of the forestry reforms came to an end (*China Forestry Yearbook* 1987). On average, each household obtained about 0.5 ha. The land area in family plots continued to expand for the next several years as collectives in most provinces in the South and Southwest reallocated forestland, often at the expense of responsibility hills. Family plots accounted for 59% of the collective forestland in 16 provinces by 1987 (*China Forestry Yearbook* 1989).

The proportion of family plots has remained steady since then. Households in 14 of the 15 villages in our devolution study still maintained family plots in 1999. (Households in the fifteenth village had returned their family plots for collective management.)

Responsibility Hills

The collectives also distributed some forested lands to individual households (and groups of households in some cases) for the purpose of improving their management. These lands were called responsibility hills (*zerenshan*). As with family plots, the collective retained ownership, but it transferred some management and production responsibilities (such as forest cleaning and protection) to the households for terms of 5, 10, or 15 years. Households had no right to transfer these resources, and they exerted only limited control over the harvest and sale of the trees on them. Households and the village collectives shared the income from responsibility hill forests according to terms agreed upon by the two parties. The terms varied from village to village according to forest accessibility and the age of the trees at the time of the distribution. For example, in Nujiang Prefecture of Yunnan, households were entitled to 20% of the income from easily accessible responsibility hill forests but as much as 60–80% of the income from less accessible forests (Nujiang Prefecture Forestry Bureau 1996).

The land area in responsibility hills has declined since its initial distribution. In some provinces, responsibility hills were incorporated into family plots immediately after their distribution from the collective (Compiling Board of Hubei 1989; Compiling Board of Guizhou 1994). In others, the local forestry cadres and other officials observed increasing rates of forest degradation and deforestation. They attributed this to poor household management, and they responded by withdrawing the responsibility hill contracts. The affected lands reverted to collective management. For example, almost all responsibility hill forests in Suining County, Hunan; the village of Wenhai, Lijiang County, Yunnan; and the villages of Eshan County, Yunnan, reverted to collective management for this reason (Li 1995, Compiling Board of Suining 1997).

Despite the decline in total land area, responsibility hills are still an important form of local forest management in many counties and provinces today. No data are available on the proportions of national or provincial forests in responsibility hills, but the proportion ranged from 36 to 60% in the villages of our forest degradation study.

Collective Management

Modified forms of collective forest management remain important today despite the popularity of household management. Its role is larger in the more remote and forest-dependent areas of Yunnan, Guizhou, and Hunan.[1] Collectives still manage some forests in all 15 villages of our devolution research sample, for example, and collective management accounts for 20–50% of the total forest area in these villages. Most of it is natural forest that is less accessible for economic activity, and the management objectives tend to focus on ecological services.

Modern collective management is unlike the collective management of the 1960s and 1970s. First, decisions regarding forest management and the distribution of forest income are now open to greater input from community members.[2] Many decisions are discussed in open forums, and previous decisions are subject to critique and revision, sometimes through a voting process. The income from forests under collective management tends to be used for improving the community infrastructure or for community services such as teachers' salaries, although households generally receive small equal distributions.

Second, in some areas (e.g., Sanming Prefecture in Fujian and Jinping County in Guizhou), the collective forests were distributed to individual households in the form of monetary shares rather than as distinct plots of forestland (Sun 1992, Song et al. 1997, Liu 2001). Some see little difference between this form of shareholding and the old collective management, because the collective still manages the land (Sun 1992, Yin 1998). Yet, even the shareholding system allows villagers at least limited input to management decisions that are now generally made by a board of village trustees.

Consequences of Devolution

The devolution polices of the early 1980s had positive effects but also unanticipated and less desirable consequences in terms of forest area and stock, stand composition, and social impacts. We will identify the consequences and then discuss their underlying causes.

An Initial Decrease in Forest Stock

The standing stock on all three categories of nonstate forests declined rapidly after the distribution of some of these resources to local households.[3] In eight of the 10 provinces in the South, the dominant region for nonstate forests, the nonstate stock declined from 944 million m^3 in the 1977–1981 forest inventory to 812 million m^3 in the 1984–1988 inventory (*China Forestry Yearbook* 1990, *MOF* 1996).[4] Comparable measures are unavailable for the 1977–1981 inventory for the southwestern provinces of Yunnan and Sichuan, where nonstate forests are also important, or for the total of all nonstate forests in China. However, most observers believe that these nonstate stocks declined as well.

The evidence from our 15 villages is consistent with the official forest inventory data for southern China. In Suining County in Hunan, for example, deforestation became such a concern for the government authorities that responsibility hills were returned to collective management in 1987, only four years after their original distribution to households. The majority of the county's nonstate forestlands have remained under collective management ever since (Compiling Board of Suining 1997). The initial impact was similar, but the final disposition of land and inventory of trees was different in the village of Laofan, Yuanmou County, Yunnan. This village experienced serious deforestation immediately after the introduction of household-based agriculture in the early 1980s. The success of household-based agriculture led to demands for a comparable household-based system for forestry and a breakdown of local collective forest management. Villagers cut most of the collective's 20-year old eucalyptus trees, and the village leadership had to sell the remaining live trees to the villagers in 1983. Today, under household management, few of those trees are still standing, but some households have planted additional trees on the lands they manage for themselves. In fact, deforestation followed the redistribution of the collective forestlands in almost all of the villages in our research sample. It was not offset by planting and new growth until the late 1980s.

Reforestation and Forest Plantations

The period of rapid harvest and declining inventory in the early and mid-1980s was followed by a period of modest expansion in reforestation in the late 1980s. Both the official forest inventory and our village surveys support this contention. Nonstate plantations in the South increased from 6.8 million ha in the 1984–1988 forest inventory to 8.7 million ha in the 1989–1993 inventory; those in the Southwest increased from 1.3 million ha to 1.75 million ha over the same period (*China Forestry Yearbook* 1990, 1995). The annual rates of increase for both regions were approximately 6%, or almost double the rate of increase for state-owned forests for this period.

Table 2-2 shows the change in total forest area (natural forests plus plantations) for different groups of the villages in our study area over reference periods from 1983 to 1995. All of these villages planted trees during the late 1980s; only the species varied. The area under forest cover increased significantly in most villages.

Stand Composition

The experience of deforestation followed by reforestation had important consequences for the stand structure and species composition of the forest, as well as for stand size and the number of individual forest holdings.

Table 2-2. *Change in Forest Area in Selective Sites*

Site	*Forest area (ha)*		*Change (%)*
Dicha, Jinping County, Guizhou	504 (1983)	1,621 (1995)	+222[a]
Datu and Shuiwei, Libo County, Guizhou	152 (1984)	1,064 (1994)	+600[b]
Suining County, Hunan	164,418 (1983)	179,707 (1990)	+9[a]

Note: Numbers in parentheses indicate sample year.
[a]Mainly achieved through tree planting.
[b]Plantation forest only.
Source: Data collected from field surveys.

Stocking. A decline in the land area in mature forests was one direct consequence of the period of rapid harvests. Only seedlings and juvenile stands remained. With no mature forests, local farmers met local timber demands with selective harvests from immature stands. This practice degraded the remaining forest even further.

The stocking of the nonstate forests averaged 40 m^3/ha in the 10 southern provinces and 75 m^3/ha in Yunnan and Sichuan for the 1984–1988 forest inventory. This is much lower than the stocking of nonstate forests in the same regions for the 1977–1981 inventory or for the state forests in these two regions in the 1984–1988 inventory. (The stocking of state-owned forests in the 1984–1988 inventory was 70 m^3/ha for the South and 197 m^3/ha for the Southwest, and it was increasing in both regions.) Stocking continued to decline in the nonstate forests in both regions from 1989 to 1993, and growth was restricted to the slower juvenile growth characteristic of the newest young plantations (*China Forestry Yearbook* 1990, 1995; MOF 1996).

The trends at the village level were similar. All mature stands within Gaohong township in Lin'an County in Zhejiang, were harvested, and the standing volume of young stands was very low in the villages of Shangfeng (14 m^3/ha), Chenjiakan (21 m^3/ha), and Hongqiao (24 m^3/ha). Local farmers believe that the forest stock in 2000 was less than half of what it was in the early 1980s. In the village of Dicha, Jinping County, Guizhou, stocking on the nonstate forests declined from 42 m^3/ha in 1983 to 31 m^3/ha in 1995.

Species Composition. The expansion of single-species stands was a second consequence of devolution. When mixed-species natural forests were harvested, they often were replaced with plantations. The new plantation forests in Gaohong, Lin'an County, Zhejiang, are composed of Masson pine, Chinese fir, or bamboo. In Huitong County, plantations of Chinese fir and of nontimber forest products such as fruit and bamboo replaced mixed forests. Single-species stands of Chinese fir are popular in Jinping County, Guizhou. When farmers select a species for planting, they select for local needs. They are unlikely to respond to broader national environmental values. Therefore, their plantations may provide fewer environmental services than the original natural forests, and they surely support less biodiversity.

Stand Fragmentation. Fragmented landholdings and forest stands were a third consequence. Although local governments and collectives dealt with the problem of land distribution for agriculture differently across the country, many separated land into units of similar productivity and distributed a unit of each productivity class to each household. This created a fragmented and scattered distribution of household plots (Chai 1997, Liu et al. 1998).

The distribution of collective forestlands generally followed the same pattern. Forests were first divided into plots according to species, age, density, site and soil quality, and distance from the village, among other criteria. These plots were redistributed such that households of similar family size received similar mixes of high- and low-quality forest, more- and less-accessible forest, and so forth (Sun 1992, Liu 2001). Consequently, most households received several tiny forest plots on different hills and different slopes. Land on one slope was often distributed among many households in one village. The average household in southern and southwestern China obtained four or five plots totaling only 1–2 ha. For example, in 1999, the average household in Zhaizao township, Guizhou, possessed 14 plots with a total area of only 4.2 ha. The largest plot in this township was 2.3 ha, and the smallest was only 0.06 ha. Also in 1999, in the village of Yixiang in Jinggu County, Yunnan, 45 ha were distributed in 169 plots that averaged 0.28 ha per plot.

The objective was to create equity among households, but the many fragmented and scattered plots increased the demands on the managerial resources of each household. Small plot size increases the difficulty and reduces the cost-effectiveness of boundary marking, protection, and most other forest operations. Since the redistribution, each household must allocate a family member to look after its many forest plots, whereas only one or two guards were necessary to protect an entire forest under collective management. Some households in Chuxiong in Yunnan—especially those with fewer members—reported that providing daily protection from theft and fire is a burden for them.

The problems of fragmented holdings are compounded by the many unclear boundaries. These are a direct consequence of the distribution process for family plots and responsibility hills in the first half of 1980s, and they remain a source of many boundary conflicts today.

Welfare and Environmental Effects

Devolution has increased household material wealth. It also has altered the flow of environmental services that local communities receive from the forest, although in an indeterminate way.

Income. Despite tremendous income growth for the country as a whole, rural poverty actually increased in the West (Gansu, Guizhou, Ningxia, Qinghai, Shaanxi, Sichuan, Tibet, Xinjiang, and Yunnan) between 1988 and 1995. Poverty also increased in mountainous areas and for national minorities (Gustafsson and Zhong 2000). Because these geographic and ethnic distinctions largely overlap many of southern China's most forested areas, they suggest that, at the very least, the devolution of forest responsibility has not succeeded in lifting poorer households out of poverty.

Nevertheless, the increased area of plantation forests represents an improvement in the wealth of local communities and farm households. It expands their sources of income. Villages in our research area reported spectacular increases in income from bamboo plantations and sometimes from rubber, fruit trees, and pine resin. Yeh (2000) reported that families harvesting matsutake mushrooms can earn up to 50,000–60,000 yuan (US$6,000–7,000) during a season, or about 10,000 yuan per capita, although 2,000–3,000 yuan per capita (US$250–350) is more common.

Income from timber varies significantly from village to village, but revenues from timber are still limited in most areas. The nonstate timber plantations, while expanding in area, are still mostly poorly stocked or immature. It will be a few more years before they begin to add significantly to the revenues of some households. Nevertheless, some farmers have been able to use their young timber stands to diversify their income opportunities. Households have used occasional harvests of their trees and forests to provide security against future needs as well as flexibility in responding to market conditions. In addition, since 1993, villages and farmers have been permitted to use their forests as collateral.

The village of Datu in Libo County, Guizhou is illustrative. Each household in the village has established about 10 ha in small plantations of Chinese fir. Each household also has a share in the collective's remaining natural forest. The plantations are valuable. One farmer has 20 ha valued at about 500,000 yuan (US$60,000) after eight years of growth. Two farmers who jointly manage their lands have 23 ha valued at 743,000 yuan (US$90,000). Farmers in Datu have used plantations as collateral for loans from commercial banks to finance animal husbandry and trucking enterprises. The village has used its collective forest as collateral to finance school construction.

Environmental Services. The impact of devolution on the environmental services of forests is ambiguous. New plantations on previously denuded land improve environmental services, but the decline of the natural forest decreases environmental services.

Our own research indicates that in some minority areas, such as among the Miao of Guizhou, greater local authority over the forest has raised the profile of environmental services in management discussions. We have received reports from sites in southern Yunnan that cultural traditions have ensured that community watershed protection take precedence over income generation from the forest. In Chapter 4, Yin will show that, subsequent to agricultural reforms, farmers in the North Central Plains aggressively planted trees to protect their agricultural lands from wind erosion.

On the other hand, Albers et al. (1998), Rozelle et al. (Chapter 6), and Sayer and Sun (Chapter 9) pointed out that the decline in standing forest volume and the general decline in natural forests, combined with the conversion to single-species plantations, suggest that the ability of China's forests to provide biodiversity, natural habitat, natural recreation, pest resistance, and carbon storage also has declined.

Explaining the Effects

This section examines the reasons why devolution policies have achieved or failed to achieve desirable results. The crucial factors are the security of household tenures; regulatory, tax, and market barriers to the effective use of forest resources; local capacity to manage forests; and problems in the allocation of family plots and responsibility hills to individual households.

Insecure Tenure

Insecure tenure is one reason why farmers harvested rapidly—as soon as they received some rights to forests in the early 1980s. It also explains, in part, their reluctance to invest resources to protect existing stands or to plant new stands (Sun 1992, Bruce et al. 1995, Albers et al. 1998).

The insecurity has three sources. The most obvious is the short term for which households have enjoyed use rights to their land. New rights to agricultural lands were initially contracted for terms of 15 years, although even this term was subject to some creative interpretation by local officials. Fifteen years is an insufficient term for many agricultural investments and Feder et al. (1992) showed that it explains the deterioration in irrigation infrastructure that occurred in the 1980s.

In forestry, family plots were allocated for long-term use, but no specific term was established. Responsibility hills were contracted for 5, 10, or 15 years. These terms are insufficient for the production of most timber species and products, and they are one explanation for the greater enthusiasm for short-maturity nontimber species such as bamboo and timber species that also produce other early-maturity forest products, such as eucalyptus leaves for medicinal oils.

Local governments often imposed a "use it or lose it" qualification on agricultural land. They imposed a similar restriction when they allocated forestlands to households (Liu et al. 1998). The collectives could take back the land when households were either unwilling or unable to plant trees on the family plots. Sometimes the collectives retained the lands they recovered as collective forests. In other cases, they redistributed the lands to other households. For example, in the study villages of Gengma and Chuxiong and in other areas of Yunnan and some southern provinces, local governments specifically identified and recovered lands that had been distributed to households in the early 1980s but had not been reforested. They subsequently leased these recovered lands to other individuals and institutions in the mid-1990s. In general, the stricter the land-use guidelines, the less secure the tenure and the more likely it is that households will harvest without reinvestment—or deforest and not invest in reforestation and management. Therefore, the government threat to recover land that has not been reforested becomes a self-fulfilling prophecy.

Finally, frequent policy changes made household managers cautious, even of good policy. During the three decades prior to the early 1980s, forest tenures—especially those of private trees—changed frequently, leading farmers to conclude that they should "get what they can while they can" (Albers et al. 1998,

Yin 1998). Major policy changes at the national level alone included the following:

- In 1956, the private forests of farm households were collectivized, but scattered trees around homesteads remained private property.
- In 1958, private trees around homesteads were also taken away by the communes.
- In 1961–1962, the rights to trees around homesteads were returned to households.
- From 1966 until the 1980s, the rights to these trees were taken away once more.

This sequence of periodic policy revision made farmers doubtful that any policy would be longstanding. It gave them more reason to harvest whenever they had the opportunity, before the policy changed again. It also diminished any incentive the farmers might have had to protect standing trees or forests or to reforest after harvesting, because they doubted that they would ever collect the return on their investment.

More recent increases in farmer participation in forest management and reforestation suggest that household confidence in forest tenures has improved. The primary policies on household forest tenures have not changed in 15 years now, contracts for both agricultural and forests lands have been renewed as their first contract periods expired, and several more recent policy changes have strengthened household rights and improved household incentives. With the increase in investor confidence, forest management and protection have increased, and an increase in forest stocks has followed. The stock of nonstate forests in the South expanded from 1,081 million m^3 in 1984–1988 to 1,136 million m^3 in 1989–1993. The stock of nonstate forests in the Southwest expanded from 778 million m^3 to 840 million m^3 for the same period (*China Forestry Yearbook* 1990, 1995).

The permission to sell live stands adds to the other incentives to manage for long-term timber production and further improved forest tenures. The authority to transfer rights to trees was first granted at the local level in Guizhou in the mid-1980s. It was granted elsewhere in Guizhou as well as in Yunnan and Fujian by the mid-1990s. The revised Forest Law of China legalized these transfers for the entire country in 1998, and farmers with willing buyers now have the authority to freely convert these assets to income. The effect of this incentive will show up most clearly in the form of additional growth in the next national forest inventory, but some local evidence already exists. Investments in tree planting and forest protection increased substantially in the village of Datu in Libo County, Guizhou, for example, in the early 1990s. The land area of forest plantations also increased in Shizong County, Yunnan, after 1993 (Zou 1994).

Nevertheless, many sources of insecurity remain despite these improvements in household forest tenure. The logging ban imposed on the upper watersheds of the Yellow and Yangtze Rivers in 1998 is one example. This ban may help control erosion and downstream sedimentation, but it surely reduces the security of tenure and the incentives for investment in the forested uplands. The expansion

Table 2-3. *Distribution of Timber Revenues (%)*

Site	Taxes	Fees	Logging, processing, and transport costs	Residual income for household or collective	Total	Year of observation
Dicha, Guizhou	51[a]		16	33	100	1998
Tiantang, Hunan	33	18	15	34	100	1998
Fangtuan, Hunan	49[a]		21	30	100	1998
Jiangle County, Fujian	77[a]			23[b]	100	1998
Jin'an County, Jiangxi[c]	60[a]			40[b]	100	1998
Hongqiao Township, Zhejiang	49[a]			51[b]	100	1997

[a]The total of tax and fees.
[b]Costs of transportation plus residual income to community or household.
[c]Bamboo timber.
Sources: Authors' field notes (1998–1999), SFA Survey Team 1999, Zhang 1999.

in the nonstate forest inventory from 1984–1988 to 1989–1993 indicates that farmers do respond to improved tenure and to a more stable policy environment, but they are cautious. Their favorable response was slow in coming and it erodes easily. One clear lesson from this experience is that decisionmakers must consider the effect on farmer confidence before implementing new policy.

Taxes and Fees, Regulations, and Marketing Barriers to Forestry

The low financial return is another reason farmers have been reluctant to plant trees (Bruce et al. 1995, Albers et al. 1998, Yin 1998, Zhang et al. 2000). Liu et al. described the problem in greater detail in Chapter 3. The return on investment in commercial timber species is especially constrained. The government takes 50% or more of timber sale receipts in the form of taxes and fees and the costs of logging, processing, and transportation account for another 20%. This leaves less than 30% for the household or the collective. When the costs of planting, management, and protection are taken into account, the returns to households or collective managers from timber production are small at best.

Table 2-3 shows the distribution of these costs in selected locations. Taxes and fees ranged up to 77% in Jin'an County, Jiangxi, and the responsible land managers, the collective or the household, seldom retained more than 35% of gross revenues from timber receipts, even as late as 1999. In Suining County, Hunan, the tax structure encourages farmers to prefer bamboo to timber species. Taxes and fees are lower for bamboo, and the household captures more of the eventual revenues from bamboo. This is also true in Lin'an County, Zhejiang, where farmers have focused on the bamboo production for its edible shoots. Other villages favor fruit trees, rubber, and other nontimber forest products in preference to commercial timber species for similar reasons. Still other households and villages prefer nontimber forest products (NTFPs). NTFPs accounted for 36% of nonstate plantations in the South and Southwest in the 1989–1993 forest inventory (*China Forestry Yearbook* 1995).

Various nonmonetary disincentives also discourage timber production. Quotas and permits on cutting, as well as permits for shipment and processing, limit the harvests of all wood products. When quotas are tight, permits may be unavailable and the product cannot be sold. Even when permits are available, the journey to the county forest bureau to obtain the permits is a burden for many households. (In a few places, such as Libo County, Guizhou, the burden has been eased by allowing the township to grant permits.)

Finally, state-owned timber companies had a monopoly on timber purchases from farm households until the 1990s. The monopoly restricted the households' opportunities to dispose of their wood products and kept prices low. This was another disincentive for timber production, and another reason why farmers gave priority to species such as fruit trees and bamboo, species whose markets were not so heavily controlled.

In sum, high taxes and fees, as well as various regulations and market barriers, remain constraints on the development of the forestry sector. Future reforms in these areas would allow the sector to grow, and they could have effects as favorable as those gained from transferring tenure to the households in the early 1980s.

Management Capacity and the Role of Government

The knowledge of forest management and the capacity to pursue management opportunities also have affected the success of devolution. These factors are functions of the accessibility of technical services and market information and the availability of financial resources. Differences in household capacity with respect to these resources explain some of the differences in forest management between households and villages.

Technical Assistance. Albers et al. (1998) argued that tenure reform alone is insufficient to induce participation in tree planting and management; public investment is necessary as well. Our fieldwork supports this perspective, particularly in poorer counties. Once capital was provided, farmers responded well to the opportunities associated with forest plantations. For example, the level of interest in planting trees in Libo Country, Guizhou, remained low until the Ministry of Forestry (now the State Forest Administration) began to finance afforestation in 1989–1993 as part of a poverty-alleviation program. The ministry provided an interest-subsidized loan of 910,000 yuan to the County Forest Bureau. The Bureau redistributed to loan to farm households as cash and seedlings. The Bureau was in charge of planning, and its representative at the township level (forestry station) provided services such as farmer training, technical assistance, and seedlings. Households contracted to plant 2,340 ha of Chinese fir and to repay the loan when their timber matured in 15 years.

Why didn't local villagers establish plantations before this time? Farmers claimed, during our interviews, that they had been limited by shortages of both capital and technical skills. In contrast with the Libo County example, the development of household forest management was much slower in counties with limited government financial and technical assistance (Chuxiong County, Yunnan, and the Shanxin Villages, Canwu County, Guangxi).

We expect an interval between a household's receipt of information about forest management practices and its action based on this information—just as there is an interval between policy reform and a farmer's sense of security that the reform will be permanent. Most farmers had been excluded from forest management decisions for the previous generation. Their previous land management experience was in a substantially different ecological, regulatory, and market context. They needed to establish experience in the newer context before expending household resources on forest management. The encouraging note, however, is that we can anticipate that household forest management will improve in the coming years as farmers build the management capacity that was largely absent when the devolution policies were first introduced. Government technical assistance can help decrease the interval over which they build this capacity.

Governance. The local capacity to govern must complement local technical and financial capacity. The local organization of the Miao people of Libo County provides an example. Libo County did not experience the deforestation common to other parts of China during the first half of the 1980s. Forest guards are generally unnecessary in Libo County, and the support for natural forest regeneration is strong because the Miao people believe that forests protect their ancestors. They do not cut trees from another person's property without permission. Where violations of the local norms occur, the Miao chief plays an active role in bringing social pressure on villagers and settling disputes.

The capacity to govern includes the ability and opportunity to develop new institutions to complement those that already exist. Some institutions have been created specifically in response to particular problems of local forest management. The locally self-initiated shareholding systems that evolved to address the problems of dispersed and fragmented forest landholdings are one example (Liu 2001). These shareholding systems accumulate numerous small and fragmented plots into units that are better suited for plantation management, and they arrange for the closure of mountain plots where natural generation (*fenshanyulin*) is preferable. Local farmers with shares in these systems participate in setting the terms of the contracts for labor inputs and the distribution of rewards from production. Where external assistance is available, these shareholding systems often draw on state or private expertise and capital that are unavailable to individual farmers. In contrast with the shareholding systems initiated by the state, these self-initiated systems build on local capacity and take advantage of local incentives (Yin 1998). Only a few such systems exist thus far, mostly in Guizhou and Zhejiang Provinces, and some observers are concerned about the members' capacity to negotiate effectively with each other and with outside partners. Nevertheless, these systems are promising means for addressing the principle shortcomings of the policy reforms to date without compromising the new set of incentives.

In another example, Yeh (2000) reported that matsutake mushroom harvesters have designed village compacts (*xiangguiminyue*) to limit access to the forest and to regulate harvest practices. These village compacts sometimes have ambiguous relationships with the state, and outsiders frequently challenge them. Therefore, the state may have a role in recognizing and perhaps even nurturing these and other local institutions that develop in response to locally perceived needs. In any event, forest management has been especially effective in the places where we observe locally initiated institutions.

Policy Implementation: Government Performance and Accountability. Household capacity varies with a host of factors. Some of them raise questions of fairness that can be addressed only with institutions that are accountable to the local population.

The process of initial land redistribution to households is at the core of many questions of fairness. The inexperience of many public officials and the brief time they had to complete their redistributive task partly account for the problems of fragmented forest holdings and subsequent boundary disputes. Hundreds of officials in each county worked on the distribution of family plots and responsibility hills. Many had no experience with cadastral surveys. They were often assisted by youths who also lacked experience. These cadres were required to complete their task in a short time—usually a few months. In the worst cases, boundaries were created as desk exercises, without accompanying field checks.

Fragmented holdings and boundary disputes were common. Those farmers who were better-educated were more aware of marketing opportunities, shareholding systems, and the arrangements for land transfers. They used this advantage to improve their positions relative to those of others with less education.

Political connections to township, county, and provincial governments helped other households to prosper. Especially in the village of Gengma, Yunnan, better-educated local officials with ties to the county town seized opportunities to lease forestland and obtain former swidden land for enterprise development, often at the expense of villagers from minority ethnic groups. When minority villagers began negotiating contracts, they often began with information disadvantages, and they gave up substantial tracts of forestland on promises of payments that never materialized.

In other counties, better-educated and better-connected villagers simply got a head start on their neighbors in obtaining forest-use rights, buying timber stands, and otherwise accumulating forest assets. Accountable village governance has proven effective in addressing such inequities. For example, the neighbors of one wealthy farmer from Suining County in Hunan, with the assistance of the village administration, pressured that farmer to return a portion of the 200 ha that he had acquired.

Since the reforms, incomes have become more unequal, particularly in western areas that are officially designated as poor (e.g., Gustafsson and Zhong 2000). Households in some villages now possess more than 30 ha, whereas others in the same villages possess only two or three hectares. The difference appears to be associated with education levels and access to information about government policies and markets. The households that got an early start with plantation development enjoy advantages today. Even for poorer households, the benefits of household management have been large enough to obtain broad support for the policy. Nevertheless, increasing differences in household wealth are a source of concern among villagers and government officials alike.

Conclusions

The devolution policies of the early 1980s improved the prospects for local forest management. Tenure reform improved the incentives for forest management, a stable policy environment engendered confidence, and household capacity has risen. Farmers now take a more active role in forest management decisions, and in many areas farmers have begun protecting existing forests and they have begun planting trees. Farm incomes have risen.

However, the results are not as positive as they could be. A legacy of high taxes, fees, and regulations remains; and the new logging ban reminds farmers that a government that has given rights and then withdrawn them once can do it again. A return to "use it or lose it" behavior is always possible as long as this uncertainty remains.

A second lesson of China's experience is that the effectiveness of any policy depends on its implementation. China's experience with the initial redistribution of forest resources in the early 1980s shows that rapid and inexperienced administration creates unintended costs—even if the policy is good. In China's case, the unintended costs were associated with managing small and fragmented parcels of land and with resolving subsequent boundary conflicts.

A third lesson is that property rights are important, but improved property rights are only one source of the improvements in forest management. These rights must be supported by an effective incentive structure, including full rights to harvest and market forest products, a generous scheme for returning the benefits to those who made the investments and took the risks, and a consistent and reliable policy environment.

The improved incentive structure must be accompanied by an improvement in local capacity to manage. The government can play a role in supplying households and communities with capital to undertake forestry operations. It can provide technical support for forestry operations, particularly in the early years as farmers take on forest management responsibilities that were previously unknown to them. Finally, it can help those who hold the new property rights to build the capacity to govern themselves. In part, this means being able to develop and enforce local norms of forest access and use. The Chinese government has had some success in this regard—first nurturing the household management system, then nurturing the shareholding system that began in Anhui, Fujian, and Guizhou. Another aspect of improved farmer capacity is the ability to negotiate with outside parties over the terms of contracts and agreements. The worst abuses of policy reforms have taken place where villagers entered into contracts with outsiders on very poor terms or with little hope of redress if the outsiders failed to honor the terms of the contract.

We recommend that governments give special attention to self-initiated shareholding systems. These systems address some of the most important technical and financial problems associated with household management while preserving the incentives that have made farmers more enthusiastic about forestry during the past two decades.

Notes

1. Liu et al. 1998 discuss the variation in agricultural reforms across China. They argue that more radical change and individualized tenure occurs in areas where the off-farm employment opportunities are greater, land is more abundant, and grain quotas are lower (and the government is less likely to fear the effects of more radical reform on national grain supply).

2. See O'Brien and Li 2000, Pastor and Tan 2000, and Oi and Rozelle 2000 for general discussions of village level democracy in China.

3. All three categories of nonstate forestlands remain the property of the collectives and all three appear together as "collective forest" in the national forest inventory. Only their management responsibilities have changed. This creates a problem in nomenclature because one of the three categories remains collective in both ownership and management responsibility. Hereafter, we refer to the three categories together as "nonstate" forests to be clear about management responsibilities and also to clarify that we mean the sum of all three categories.

4. Data are unavailable for Guangxi and Guizhou Provinces, which had lower levels of standing nonstate forest volume.

References

Agarwall, Arun, and Jesse Ribot. 2000. *Decentralization, Participation and Accountability in Sahelian Forestry: Legal Instruments of Political-Administrative Control*. Cambridge, MA: Harvard University, Center for Population and Development Studies.

Albers, Heidi J., Scott D. Rozelle, and Li Guo. 1998. China's Forests under Economic Reform: Timber Supplies, Environmental Protection, and Rural Resource Access. *Contemporary Economic Policy* 16: 22–33.

Bruce, J.W., S. Rudrappa, and L. Zongmin. 1995. Experimenting with Approaches to Common Property Forestry in China. *Unasylva* 46(180): 44–48.

Central Committee of Communist Party of China and State Council. 1980. Directives on Making Strenuous Efforts to Plant Trees and Develop Plantations. Reprinted in *China Agricultural Yearbook*, 1981 [in Chinese]. Beijing: China Agricultural Press.

———. 1981. Decisions on Issues of Protecting Forests and Developing Forestry. Reprinted in *China Agricultural Yearbook*, 1982 [in Chinese]. Beijing: China Agricultural Press.

Chai, Joseph. 1997. *China: Transition to a Market Economy*. Oxford, U.K.: Oxford University Press.

China Forestry Yearbook. 1987, 1989, 1990, and 1995. [In Chinese]. Beijing: China Forestry Press.

Compiling Board of Guizhou. 1994. *Guizhou Provincial Annals of Forestry* [in Chinese]. Guiyang, China: Guizhou People's Press.

Compiling Board of Hubei. 1989. *Provincial Annals of Forestry* [in Chinese]. Wuhan, China: Wuhan Publishing House.

Compiling Board of Suining. 1997. *Suining County Annals* [in Chinese]. Beijing: Publishing House for Local Annals.

Feder, Gershon, Lawrence Lau, Justin Lin, and X.P. Luo. 1992. The Determinants of Farm Investment and Residual Construction in Post-Reform China. *Economic Development and Cultural Change* 40(3): 287–312.

Gustafsson, Bjorn, and Wei Zhong. 2000. How and Why Has Poverty in China Changed? A Study Based on Microdata from 1988 to 1995. *The China Quarterly* 164: 983–1006.

Li, Wenhong. 1995. Several Practices in Tightening Up Forest Management in Eshan County (in Yunnan) [in Chinese]. *Yunnan Forestry* 3: 3–13.

Liu, Dachang. 2001. Tenure and Management of Non-State Forests in China Since 1950: A Historical Review. *Environmental History* 6(2): 239–263.

Liu, Dachang, David Edmunds, and Eva Wollenberg. 2002. The Promises and Limitations of Devolution and Local Forest Management in China. Unpublished working paper. Bogor, Indonesia: Center for International Forestry Research (CIFOR).

Liu, Shouying, Michael Carter, and Yang Yao. 1998. Dimensions and Diversity of Property Rights in Rural China: Dilemmas on the Road to Further Reform. *World Development* 26(10): 1789–1806.

Manor, James. 1999. *The Political Economy of Democratic Decentralization*. Washington, DC: The World Bank.

Menzies, Nicholas, and Nancy Peluso. 1991. Rights of Access to Upland Forest Resources in Southwest China. *Journal of World Forest Resource Management* 6(2): 1–20.

MOF (Ministry of Forestry). 1996. *A Brief Account of Contemporary China's Forest Resources* [in Chinese]. Beijing: China Forestry Press.

Nujiang Prefecture Forestry Bureau. 1996. *Nujiang Prefecture Annals of Forestry* [in Chinese]. Kunming, China: Yunnan Ethnic Publishing House.

O'Brien, Kevin, and Lianjiang Li. 2000. Accommodating "Democracy" in a One-Part State: Introducing Village Elections in China. *The China Quarterly* 162: 465–489.

Oi, Jean, and Scott Rozelle. 2000. Elections and Power: The Locus of Decision-Making in Chinese Villages. *The China Quarterly* 162: 513–539.

Pastor, Robert, and Qingshan Tan. 2000. The Meaning of China's Village Elections. *The China Quarterly* 162: 490–512.

Rondinelli, D. 1981. Government Decentralization in Comparative Perspective: Theory and Practice in Developing Countries. *International Review of Administrative Science* 47(29): 133–45.

Rondinelli, D., J. McCullough, and R. Johnson. 1989. Analysing Decentralization Policies in Developing Countries: A Political-Economy Framework. *Development and Change* 20: 57–87.

SFA (State Forestry Administration) Survey Team. 1999. A Study on Timber Taxes and Fees: Hunan Case. *Forestry Economics* (China) 117: 26–35.

Song, Yajie, William Burch Jr., Gordon Geballe, and Liping Geng. 1997. New Organizational Strategy for Managing the Forests of Southeast China: The Shareholding Integrated Forest Tenure (SHIFT) System. *Forest Ecology and Management* 9(1): 183–194.

Sun, Changjin. 1992. Community Forestry in South China. *Journal of Forestry* 90(6): 35–40.

Yeh, Emily. 2000. Forest Claims, Conflicts, and Commodification: The Political Ecology of Tibetan Mushroom-Harvesting Villages in Yunnan Province, China. *The China Quarterly* 161: 264–278.

Yin, Runsheng. 1998. Forestry and the Environment in China: The Current Situation and Strategic Choices. *World Development* 26(12): 2153–2167.

Yin, Runsheng, and David Newman. 1997. Impacts of Rural Reforms: The Case of the Chinese Forest Sector. *Environment and Development Economics* 2(3): 289–303.

Zhang, Xiaojing. 1999. The Forestry Taxation and Charges Issue in the Collective Forestry Area of Fujian and Jiangxi Provinces. *Forestry Economics* (China) 6: 36–45.

Zhang, Yaoqi, Jari Kuuluvainen, and Jussi Uusivouri. 2000. Impacts of Economic Reforms on Rural Forestry in China. *Forest Policy and Economics* 1: 27–40.

Zou, Wenhong. 1994. A Good Policy Resulted in the Greening of More Than 1,000 ha of Denuded Forestland [in Chinese]. *Yunnan Daily*, Dec. 19, 7.

CHAPTER 3

Taxes and Fees in the Southern Collective Forest Region

Jinlong Liu and Natasha Landell-Mills

CHINA'S FOREST TAXES are high, and its system of taxes and fees on forestlands and forest products is complex. These are disincentives for forest management, and both household welfare and the environment are worse off because of them. The continuing development of the forestry sector and of the regions that rely on the sector depends on a reduction in these taxes and fees and a simplification of the system of assessment and enforcement.

Or such is the argument, in any case. Liu and Edmunds agree with it in Chapter 2. Others have made the same charge (Bruce et al. 1995; Albers et al. 1998; Yin 1998; Li 1999; Chen et al. 1999, 2000; Zhang 1999; Zhang 2001). The argument, however, is only a contention without a better sense of the evidence that supports or rejects it. Indeed, if there is supporting evidence, then one still needs to ask whether the problem is as great as some seem to think. The literature available to audiences who read English is undeveloped beyond this simple contention.

Our objective in this chapter is to provide the evidence. We will examine the structure of taxes and fees for forestry for one important region in China. Our fundamental conclusion will be that the full collection of taxes and fees is large, indeed—often greater than 50% of gross revenues—although it also varies widely across counties and provinces. We agree that the current system of taxes and fees is a serious disincentive for improved forest management and that it has negative social and environmental effects as well.

Our evidence is from China's South, from Hunan and Jiangxi Provinces in particular. The South is one of three major regions for commercial forestry, and it is the region best known for collective forest management. It contained 85% of

The material in this chapter is part of a larger assessment undertaken by the International Institute of Economic Development. See Liu et al. 2001 and Lu and Landell-Mills 2002.

China's collective forests in 1998, and 80% of all forestland in the South is under collective management. Collective forests are more responsive to market criteria (and to market disincentives such as taxes and fees) than the state-owned forests that dominate some other regions.

The sources of our information on Hunan and Jiangxi are accounts maintained by the county forest authorities and original surveys from three counties in each province. The counties represent a range of landscape and forest productivity types: a mountainous area of primary natural forests (Huaihua Prefecture, Hunan, and Xiushui County, Jiangxi), a hilly area where agriculture predominates (Yueyang County, Hunan, and Yingxiu County, Jiangxi), and an area with both characteristics and some competition for land use between commercial forestry and agriculture (Lingxing County, Hunan, and Fengyi County, Jiangxi).

The chapter begins with a review of the existing system of taxes and fees, including an itemized list of the taxes and tax rates in the six counties and two provinces. Subsequent sections of the chapter discuss the efficiency of collection and the allocation of revenues from taxes and fees, and then summarize the economic, social, and environmental impacts of this system of forest taxation. A final section contains our recommendations for reform.

The Current System of Taxes and Fees

China followed a top-down model of economic planning prior to the 1980s. Production, marketing, and consumption were all part of a unified central system that included the authority to collect all government revenues and allocate all expenditures. A period of general reform and decentralization followed, but the central government's system of taxation and expenditure remained unchanged. Its agencies were overstaffed. Deficits occurred at all levels of government throughout the 1980s and 1990s, and they still exist today.

The State Council began addressing these problems in 1984 by approving a series of policies granting the agencies of the central and local governments the authority to collect their own taxes and fees. Neither has the authority to establish new taxes, and the taxes that the central authority allows them do not provide sufficient revenues for their operations. Therefore, both have developed fee systems to collect the additional revenues. A few local governments have commercialized some services and the collection of revenues to pay for them. Over time, the agencies of the central government have created more and more fees, and the local governments have developed extensive lists of many small and dispersed charges.

In the overall system in 2001, only the State Council has the authority to legislate taxes, and only the Central People's Congress has the authority to approve charges or fees. There are three categories of taxes—reflecting three categories of ownership of the revenues:

- state taxes, including consumption taxes, customs duties, and a consumption tax collected by customs;

- local taxes, including agricultural taxes, education fees, and taxes for urban construction and maintenance; and
- shared taxes, including a valued-added tax shared between the state (75%) and the province (25%).

Revenues from each of these taxes support public expenditures in the sector of the central government, province, or county that corresponds to the revenue source.

In addition to these taxes, the provincial people's congresses can approve various local charges, and local authorities often add their own unofficial charges. Since 1996, the revenues from officially approved charges have been placed in accounts held by the finance sector for subsequent dispersal by the appropriate agency. The forestry authorities at the central, provincial, and county levels apply to the finance authorities at the same level for approval to use these funds.

Forest Taxes and Charges

The State Council formulated the current system of forest taxes and charges as part of the fiscal reform of 1994. The following taxes and charges are levied on forestland, forests, and forest products from plantations and natural forests, whether state-owned or collective, unless specifically indicated otherwise.

Special Agricultural Products Tax. Local authorities have the discretion to levy a special agricultural products tax (SAPT) on income earned from forest products. Revenues from this tax become income for the local government that levies it.

The tax is restricted to rates of 8% of income from logs, bamboo cane, natural resin, and tree oil and 10% of income from lacquer and natural rubber. It is collected from producers at the farm gate and again from buyers at the market, which means that the tax is levied twice. Taxable income is calculated as the yield or quantity sold times either the government procurement price or the market price, whichever is locally relevant. The tax is levied on the date of sale and is payable to the local fiscal authority within 30 days.

Reductions and exemptions from this tax apply for forest products used in research, for production during the first three years of newly reforested wasteland, for producers affected by natural calamity, and for producers defined as poor by national provision.

Other Taxes Affecting Forestry.

- Value-added tax: The state fiscal authority collects taxes on a percentage of the value added in production. For enterprises with products or services valued at more than one million yuan/year or with retail sales or wholesale products valued at more than 1.8 million yuan/year, the rate for logs and bamboo cane is 13% of the sale value minus 10% of raw material costs. The rate for products processed from wood or cane is 17% of the sale value minus 10% of raw material costs. Small-scale operators and nonprofit institutions pay a general rate of 4% of the sale value. The central and local governments share revenues from this tax in three-to-one portions.

- Education value-added tax: The local fiscal authority collects an additional 3% of the value added from enterprises and individuals involved in the processing and sale of timber and bamboo. Local governments manage these revenues for education.
- Urban construction and maintenance tax: The local fiscal authority collects yet another share of the value added from enterprises and individuals involved in the processing and sale of timber and bamboo, and allocates these revenues for use in urban maintenance and construction. The rate for this tax is 7% for cities, 5% for counties and townships, and 1% for others.
- Income tax: All enterprises pay a tax of 33% on income. The central government collects the full tax from all enterprises but reimburses all except the first 5% on products made from the residues of processing or harvesting or from low-quality wood. (In actual application, only state-owned enterprises receive this reimbursement. Private enterprises and joint ventures do not.) Foreign investors and foreign enterprises can obtain a reduction to 15–30% of the normal income tax for 10 years upon application to the fiscal authority of the State Council. The income of township forestry technical stations is exempt, as is income from forest harvests.[1]

Forest Charges Approved by the State Forest Administration. The State Forest Administration (SFA) has approved four fees that apply throughout the country.

- Afforestation fee: The local forest authority collects 12–15% of sale revenues from logs and bamboo to pay for afforestation, technical assistance, and many forest practices.
- Maintenance and upgrade fee: The local forest authority collects 8–10% of sale revenues from logs and bamboo to cover the depreciation of fixed capital in state forest harvest enterprises. The income from this fee is intended for updating equipment and constructing forest roads.
- Forest protection and construction fee: The local forest authority collects 5 yuan/m^3 of lumber and generally 0.2 yuan per bamboo cane to pay for the costs of policy implementation, forest fire protection, and road construction. Timber operations and enterprises approved for timber harvesting pay this fee. Rural collectives and farmers are exempt.
- Forest quarantine fee: The local forest authority collects 0.2% of the sale value of timber, seeds, and bamboo and 0.8% of the sale value of seedlings to pay for education, technician training, labor and equipment costs, and other items related to the control or treatment of quarantined items.

In practice, the various local authorities share all of these charges. Revenues from all four tend to be lumped together and allocated to various forest activities according to the preference of the agency making the transfer.

Forest Charges Approved by the Provincial People's Congresses. The people's congress for each province can approve fees. The following fees have been approved in provinces within the Southern Collective Forest Region (SCFR). Other fees may also exist in some provinces.

- Forest restoration fee: The provincial forestry agencies collect 3% of the sale price of timber and bamboo products in Fujian and 5% in Guizhou to pay for reforestation.
- Insect and disease control fee: The Jiangxi provincial forestry agency collects 4–5 yuan/m^3 of timber and 0.2 yuan/m^3 of bamboo to pay for insect and disease control.
- Fire protection fee: The Jiangxi provincial forestry agency collects 2 yuan/m^3 of logs (measured at the farm gate) for forest fire control.
- Administration fee: The Jiangxi provincial forestry agency collects 0.6 yuan/m^3 of timber and 0.05 yuan/m^3 of bamboo cane for its administrative costs.

Unofficial Forest Charges. Agencies below the provincial level are not authorized to impose fees other than those already listed. Nevertheless, many unauthorized charges occur within the SCFR and they vary greatly between localities. In Jiangxi, for example, more than 30 such charges are levied for the design of harvest plots, construction of infrastructure, forestry-sector administration, flood prevention, electric power development fund, biogas development, and so on. The fiscal, commercial and industry, education, water conservation, and power development sectors, as well as the forestry agencies, levy these fees.

Avoidance and Enforcement

Avoidance and enforcement are serious problems in many counties. Avoidance is as high as 50% for the SAPTs and perhaps as high as 90% for the combination of afforestation, maintenance, and construction fees.

Table 3-1 illustrates this issue with evidence from the official accounts of five forest enterprises. In four of five enterprises, the official accounts show payments of taxes and charges in the range of 8 to 41% of revenues, a range that is less than the official local fees. Most enterprises maintain two sets of accounts and the second (unofficial) accounts will be a closer reflection of actual income and expenditures. We anticipate that an examination of the second set of books would reveal an even greater level of tax avoidance.

Variations and Innovations in These Taxes and Fees

This system of taxes and fees is not static. As with China's reforms in other sectors, various levels of forest authorities have experimented with their own modifications. The SFA modified its policy on forest charges in 2000. The new policy allows provincial forest authorities to reduce SFA-approved charges by a set percentage as an incentive for foreign investment. Only Guangdong has responded to date. In Guangdong, 50–70% of the fees for afforestation and for maintenance and use can be returned to foreign investors after their enterprises successfully replant a harvested area. Up to 85% of the forest protection and construction fee can be returned when the foreign enterprises afforest more than 33,000 ha to support a pulp-making facility or other facility producing fine wood products.

Table 3-1. *Functional Burden of Taxes and Charges*

Village	*Qunhe*	*Rushi*	*Rushi*	*Qingshi*	*Fangshan State Forest Farm*
Township	Lujiao	Wenbai	Wenbai	Yunxi	
County	Yueyang	Lingxiang	Lingxiang	Yuexi	Fengyi
Province	Hunan	Hunan	Hunan	Hunan	Jiangxi
Forest manager	Village collective	Village collective	Private	Private	State
Area (ha)	15.6	135.6	0.8	43.3	1,760
Forest revenues, including taxation and charges (yuan)	3,190	34,000	>400	12,000	276,000
Cost for forest guard (yuan)	0	3,700	0	0	NA
Net income/ha (yuan)	204.5	220.7	500	257.1	156.9
Share of taxes and charges (%)	41.5	29.4	38.1	8.5	66.7

Source: Records of county forest authorities.

Other provinces and local authorities have attempted to reduce their forest charges. In Jiangxi, for example, the forest authority reduced the official base price for calculating the revenues that are subject to forest charges. In a few counties, the Jiangxi forest authority is testing the effect of a 50% reduction in charges on the products of thinnings from juvenile and immature stands. Its objective is to improve the incentive to manage juvenile and immature forest stands. In another example, the Huaihua Prefecture, Hunan, has tried to eliminate all unofficial forest charges.

These efforts have not fared well. In Jiangxi, prefecture authorities commonly set the official base price for calculating forest charges at a rate that is 10% higher than the base price established by the provincial forest department, and county authorities often raise that base price another 20%. Of course, raising the base price only offsets a proportional decrease in the tax rate. In Huaihua Prefecture, the number of unofficial forest charges initially declined, but a longer and more burdensome list of charges subsequently took its place.

Meanwhile, other variations in the system of taxes and fees have emerged without official central government approval. In Yueyang County, Hunan, for example, private enterprises do most of the logging. They contract with forest owners and then negotiate their payments of forest charges with the forest authority.

The Current Government Reform Program

The central government has begun testing a reform program for all rural revenues and expenditures in Anhui (and two or three counties in other provinces). Its long-term objective is to standardize the system of rural taxation and reduce the

Table 3-2. *Forest Taxes and Charges as Percentage Shares of Log Value at the Roadside, by Geographic Condition and Political Location*

	Mountain forest region		*Mixed forest-agricultural region*		*Hilly agricultural region*	
	Huaihua, Hunan	*Xiushui, Jiangxi*	*Lingxiang, Hunan*	*Fengyi, Jiangxi*	*Yueyang, Hunan*	*Yongxiu, Jiangxi*
Taxes	25.3	20.8	16.0	24.2	16.0	17.6
Approved charges	23.3	29.6	26.0	30.9	21.8	29.0
Unofficial charges	2.4	6.5	2.5	6.2	0	10.6
Total	51.0	56.9	44.5	61.3	37.8	57.2

Source: Records of county forest authorities.

fiscal burden in all of China. The plan is to eliminate many charges and fees and to streamline both the entire revenue system and the delivery of government services. Forestry is part of agriculture in the Anhui pilot program, and the SAPT will be the primary source of revenue. It will be collected at a rate of 8–10% at only one point (instead of the current double collection). Village authorities will be permitted to collect an additional amount equal to 20% of the tax for local use.

Characterizing the Full System

We reviewed the records of six county authorities in three characteristic regions of forest activity in Hunan and Jiangxi in order to assess the effect of taxes and other charges on the forest sectors of those regions. The regions were selected to illustrate the different levels of sectoral importance to the local economy: mountainous regions where forestry dominates, hilly regions where agriculture dominates, and regions of mixed agricultural and forestry importance. Table 3-2 summarizes our observations.

In general, approved taxes and fees absorbed 42–50% of the sale value of logs and bamboo cane. Unofficial charges absorbed up to 10% more. The forest manager was left with between 43 and 62% of the revenues from log or bamboo sales to cover all the costs of forest management and logging.

The variation among sample jurisdictions was largely due to differences in the set minimum base price for the tax calculation. Nevertheless, other differences are also apparent. Not all local governments collected at the same stage of the production process, but all tried to collect as much as possible as early as possible. In some cases, local authorities extracted taxes and charges when the forester obtained a bank loan to support the harvest operation. The certificate authorizing the right to harvest was the leverage, the government made the payment of expected taxes a precondition for obtaining the certificate, and the certificate was a precondition for obtaining the loan. Other local authorities did not allow the deduction of silvicultural costs from income taxes. In still other cases, local authorities contracted with forest or timber processing enterprises to pay a fixed amount of income and value-added tax, regardless of the profitability of the forestry activity.

The differences between jurisdictions are substantial, but the overall rate of government collection was generally greater in the mountainous region. The forestry sector is generally more important to the local economy of this region than the other two. The relatively larger size of the sector in this region probably induces greater government attention and surely indicates that the forestry sector is a greater potential source of local revenues.

Efficiency in Collection and Use

Inefficient Collection

Two factors underlie the inefficiency of revenue collection: the act of collecting is costly relative to the total revenue recovered and, even then, total recovery is only a limited share of the fully authorized level of collection.

Many of the various taxes and charges are imposed on the same product, but they are collected by different agencies, creating a redundant system of collection, excess bureaucratic employment, and duplicated recordkeeping. In some counties, more than half of the public sector's forestry personnel are employed in revenue collection.

Furthermore, in Hunan and Jiangxi, only 40–70% of all authorized revenues are collected. No revenues are collected on illegal harvests, yet in some counties in Jiangxi, total harvests (including illegal harvests) were 50% higher than the approved harvest level. A few log dealers for township collectives and primary processing enterprises consumed 6–10 times their approved harvest levels. Government revenues would rise sharply with better control over illegal harvests and, therefore, more complete taxation of actual harvests.

Some of the losses in government revenues were due to negotiated payments that were lower than the official taxes and charges would have been. But taxes that are so high they would have made the sale untenable in the first place are one reason for some negotiated rates. From the government's perspective, the negotiated revenues are better than no revenues at all. In other cases, the losses in government revenues were due to corruption and illegal marketing. However, it should be clear that high taxes and charges were one reason for these losses as well. The higher the taxes and charges, the greater the incentive to avoid them.

It is conceivable that, with lower taxes and charges, the incentive for avoidance would be lower, the compliance rate would be higher, and the total collection of government revenues would be greater. With better compliance, the costs of monitoring and enforcement would be lower, and the net revenues available for pursuing the real objectives of government management would be greater. Of course, these are unexamined issues, but they merit careful scrutiny if China desires to improve the economic performance of its forestry sector.

Misuse of Revenues

The revenues from official forest charges are all intended for forest development. In fact, a very large share of these revenues is allocated for salaries and operational costs in the forestry agencies. The central government has tried to correct the problem of overstaffed agencies, but each new reform seems to bring new responsibilities and additional staff to the agencies. For example, the administrative role of the local forest authorities has recently expanded to support the new environmental responsibilities associated with the Natural Forest Protection Program of 1998.

Other revenues have been transferred outside the forestry sector, and some have been transferred away from impoverished rural areas to support urban services. For example, in Xiushui County, Jiangxi, 94% of the revenue from forest charges went to afforestation in 1992. By 1999 only 33% went to afforestation and other activities in the forestry sector.

Impacts

Economic

High taxes and charges decrease the net value to the land manager for any forest activity. As they decrease this value, high taxes also decrease the demand for and the value of forestland. We can observe these effects in different ways on collective and state forestlands:

- from local trends in prices, taxes, and charges; the relative trend in the land area devoted to forests; and the trend in forestland values for collective forests, and
- from trends in the comparable profitability of the state-owned forest enterprises (SOFEs).

In Jinzhou County, Hunan, the price of pine logs rose from 360 yuan/m^3 in 1987 to 555 yuan/m^3 in 1999. Meanwhile, taxes and charges rose from 152 yuan/m^3 to 342 yuan/m^3. The net gain to forestry was 5 yuan/m^3, or less than 1.5% over the 12-year period. In contrast, during the same period rice prices also rose approximately 60% but agricultural taxes and charges officially remained unchanged. We would expect the conversion of forests to cropland under such circumstances. In fact, some forestlands were converted to farmland, and in many places the extent of conversion was dramatic. In one of our six survey counties, Yongxiu County in Jiangxi, forestland lost 70% of its lease value between 1987 and 1999. In Yongxiu, farmers established 3,000 ha of new plantations in 1987, but high taxes and charges reduced that to only 549 ha of new plantations in 1999.

The evidence is a little different for the SOFEs. These enterprises cannot sell their lands or otherwise transfer lands out of forestry. For the SOFEs, higher taxes and charges would be reflected in decreasing profitability on the same lands. Indeed, this is what we observe. The number of unprofitable state-owned enterprises increased over time, until more than 70% were in deficit in 1999. The Fengyi Fangshan Forest Farm is one example. It was recognized as a "flagship" farm with a long history of good management and high financial returns from 1,630 ha of forest. However, by 1999, its income was six million yuan, and its expenditures included three million yuan in local taxes and one million yuan in other fees. By 1999, its operation was so unprofitable that it had accumulated three million yuan in bank debts and owed more than three million yuan in unpaid salaries to its staff.

Social

The change in Fengyi Fangshan Forest Farm operations from highly profitable to serious financial arrears was sufficiently great and rapid to prompt the farm's workers to collectively petition the government for relief and then to conduct strikes against the government. A similar sequence of events has occurred on other forest farms: high and increasing taxes, net losses to the farm, shortages in wage payments, and then social and political protest.

Taxes are high, and revenues from forest taxes are not redistributed to poorer counties and poorer households. Eighty percent of the officially designated poverty counties are in forested regions where the wages of employees of the SOFEs average only 2,300 yuan annually, one-third of the national average for all workers. Yet the tax revenues taken from forestry do not tend to support services in these poorer regions; rather, they are transferred to better-off and more urban areas. For example, in Xiushui County, Jiangxi, 46 million yuan of taxes and fees were collected between 1992 and 1998. None were returned to the forestry sector. Most of these revenues were used to run government agencies in the towns. Approximately 30 million yuan were transferred to prefecture and provincial forest authorities, and only about 16 million yuan were returned to the county. Thirty percent of the county's share was spent on salaries and staff services in 1992. Seventy percent was spent on salaries and staff services in 1998. When tax revenues are allocated to forestry, staff salaries of the forestry authorities are a primary claimant, and most of the staff of the forest authorities live in towns and cities.

Environmental

The environmental impacts of high forest taxes and charges are mixed. High taxes and charges make harvesting less profitable. As they make harvesting less profitable, they slow the rate of destruction of the natural environments and diminish the rate of losses in environmental services such as biodiversity, thereby creating environmental gains. On the other hand, high taxes and charges are disincentives for managing existing forests and for reforestation and, therefore, disincentives for the environmental gains that managed forests provide.

Where taxes and fees are high, existing forests (both natural stands and plantations) are drawdown without subsequent reinvestment. In Chapter 2, Liu and Edmunds explained that this was the general experience throughout China in the early 1980s. However, they also described the extension of forest management that occurred in the late 1980s and 1990s—while taxes and fees were rising. In Chapter 4, Yin will also observe reforestation due to improved tenure and largely for the purpose of satisfying nonmarket environmental objectives.

How can we reconcile these observations with our expectations that higher taxes and charges are disincentives for forest management? We can hypothesize that improved tenure and high taxes and fees have had opposite effects on forest management and the extent of forest cover. Where forest cover has increased, the dominant factors have been the combination of improved tenure and the local importance of nonmarket forest values or marketed forest products that have not been subject to as many restrictive taxes and charges (products such as bamboo, fruit, and rubber). However, where forests are more plentiful, the beneficial effects of nonmarket values (e.g., erosion control) are not received locally, and crops like bamboo and fruit are not as important, we expect that the effects of high taxes and fees will dominate and that forest cover and forest-based environmental services will decline.

This hypothesis predicts a slower rate of decline in forest cover in regions of dense natural forest and a more rapid decline in forest cover and forest-based environmental services in those agricultural regions that are not locally affected by severe erosion. It also predicts forest conversion to cropland in the latter regions, because high forest taxes would cause a shift in the relative returns (favoring agricultural production over marketed forest products). Of course, land conversion out of forestry implies further losses of forest-based environmental services of all varieties. Finally, the hypothesis predicts reforestation only in agricultural areas where erosion has serious local effects and in regions where horticultural, rubber, or bamboo crops are important because these forest products are not affected by the high taxes on marketed timber.

The SCFR exhibits some support for these predictions. The natural forest cover has declined in the mountainous forest areas of our observations (identified in Table 3-2), although the rate of decline appears to have tapered off in recent years. Forests have been converted to agriculture in the mixed forest–agriculture counties, and some limited forest management for erosion control is evident in the hilly areas where agriculture dominates and the previous tree cover was sparse.

Another set of environmental effects appears as well. If high taxes and fees are disincentives for the management and harvest of commercial forest products, then they must encourage the industries that consume wood as a raw material to seek substitute inputs. What are the environmental effects of substitution?

- Some of China's substitutes come from imported logs and other unprocessed wood. In this case, China is saving its own forest environment by transferring its environmental losses to the countries from which it imports. This scenario is to China's advantage. Whether it creates global gains or losses depends on the relative environmental effects of logging in the two countries, China and the source of China's imports.
- Some substitutes come from alternative raw materials used in construction and in paper products, the two major commercial wood products. The net environmental effect of substituting stone, brick, block, and other materials for construction wood is uncertain, but substitution in paper production creates serious new environmental losses. Agricultural residues are the dominant substitute for wood in China's paper industry. They are more damaging sources of suspended solids and biological oxygen demand than wood. Because paper

production is already the foremost source of pollution in China's rural environment, high forest taxes and charges only make a critical problem worse in this case.

In sum, the overall net environmental effects of high forest taxes and fees are uncertain, but many of the component effects are environmental losses. The Chinese government has recognized the critical importance of the environment for in its Natural Forest Protection Plan of 1998 and in its fundamental strategy for the twenty-first century.[2] (Some of these issues are further considered in Chapters 8 and 10.)

Conclusions: Exploring Options for Reform

Two fundamental questions underlie any insight into the appropriate level and focus of forest taxes and charges:

- How can the performance of the forestry sector be improved? and
- What is the role of public forestry authorities in an improved sector?

The answers to these questions are related to improved incentives for forest management. They are suggestive of the necessary modifications in the responsibilities of the public forestry authorities and the activities of their personnel.

The Anhui pilot tax reform project addresses these questions. It combines several taxes into one broad SAPT tax, eliminates many charges, and thereby reduces the overall tax burden. By reducing the tax burden, it improves the incentives for forest management. By combining taxes, it eliminates the need for multiple agencies to be involved in tax collection and eliminates the collection activity for personnel in agencies that no longer have that responsibility. One original objective of the SAPT was to provide funds to support agricultural production. China now has a food surplus, and this objective is no longer necessary. Many of the revenues from forest taxes and charges have been used to support the personnel involved in revenue collection for the various forest authorities. But revenue collection will be simpler and less expensive in the pilot project. As the efficiency of revenue collection increases, the demand for revenues decreases, the tax burden decreases, and the incentives for forest management improve.

However, the Anhui pilot project also leaves two related questions unanswered: what happens to the redundant forestry personnel, and how can the government provide for critical nonmarket, forest-based environmental services? We can gain some perspective on these two questions, by considering the broad objectives of public-sector involvement in forestry. These are, first, to assist the development of the forestry sector and, second, to ensure the protection of nonmarket environmental services. Public agencies can assist in sectoral development by expanding market reform into the forestry sector—that is, by removing their own restrictions on private harvest levels and on the shipment of forest products, increasing the rate of transfer of state-owned commercial timberlands to the private sector, improving the tenurial rights of households and collectives, and auctioning harvest rights to mature timber on the lands that remain state-owned. These reforms can be accomplished, for the most part, by simply reducing the oversight activities of the forest authorities. As the requirements for government oversight diminish, the demand for revenues to support the personnel responsible for oversight will diminish, the tax burden will be reduced further still, and the incentives for forest management will improve yet further.

Some of the cadres previously involved in revenue collection can be assigned new responsibilities in forestry extension, tasks designed to transfer knowledge about forest management and improve the rate of forestry-sector development. Others can become more actively involved in environmental protection, both transferring knowledge about good environmental forest practices and monitoring and enforcing environmental regulations. However, some personnel will be redundant even after the reassignment of responsibilities. These remaining excess personnel are an unsolved problem.

Finally, the new responsibility for good environmental management requires careful thought. Some improvement may occur with good monitoring and enforcement, but the real key is to provide the right incentives. These may mean new charges for environmental services such as forest recreation and downstream protection against flooding and sedimentation. The revenues from these charges can be used to compensate the forestry authorities that provide the recreational forests and the upstream private managers that introduce erosion control practices on their lands.

The most fundamental point is that the overall burden of taxes and charges must decrease substantially if the forestry sector is to develop. How much the burden can be reduced depends on the size of the forestry agencies and their own demands for funds to support their staff. As government decisionmakers consider this question, they must balance the potential gains from the forestry agencies' uses of funds against the loss in incentives for good private forest management that occur when these agencies collect those funds.

Notes

1. Township forestry stations provide technical assistance and nursery stock for local farmers. They may obtain income from the sale of either.

2. Premier Zhu Rongji, in the Government Report to the National People's Congress on March 5, 1999, identified sustainable development as a fundamental government strategy for the twenty-first century.

References

Albers, Heidi J., Scott D. Rozelle, and Li Guo. 1998. China's Forests under Economic Reform: Timber Supplies, Environmental Protection, and Rural Resource Access. *Contemporary Economic Policy* 16: 22–33.

Bruce, J.W., S. Rudrappa, and L. Zongmin. 1995. Experimenting with Approaches to Common Property Forestry in China. *Unasylva* 46(180): 44–48.

Chen, Xie, Guoqiang Feng, and Xianbo Wu. 1999. Investigation and Study on Forestry Taxation and Charge-Based on Hunan Province *Forestry Economics* (China) 6: 26–35.

———. 2000. Investigation and Study on Forestry Taxation and Charge-Based on Hunan Province *Forestry Economics* (China) 1: 21–28.

Li, Mingfeng. 1999. Problems and Reform Recommendations of the Forestry Taxation and Charges in Guangxi Autonomous Region. *Forestry Economics* (China) 1: 56–63.

Liu, Jinlong, Wenfa Xiao, Yannan Hou, Chen Xie, Xiaojing Zhang, Guang Wang, Xuemei Zhang, Xiaocun Huang, and Jizhong Li. 2001. Study on the Forestry Taxation and Charges System in South China Collective Forestry Region. Unpublished background paper. London: International Institute for Environment and Development.

Lu, W., and N. Landell-Mills (eds.) 2002. *Getting the Private Sector to Work for the Public Good: Instruments for Sustainable Private Sector Forestry in China*. London: International Institute for Environment and Development.

Yin, Runsheng. 1998. Forestry and the Environment in China: The Current Situation and Strategic Choices. *World Development* 26(12): 2153–2167.

Zhang, Xiaojing. 1999. The Forestry Taxation and Charges Issue in the Collective Forestry Area of Fujian and Jiangxi Provinces. *Forestry Economics* (China) 6: 36–45.

Zhang, Yaoqi. 2001. Economic Transition and Forestry Development: The Case of China. Ph.D. thesis. Helsinki, Finland: University of Helsinki, Department of Forest Economics.

CHAPTER 4

Central Characteristics of Reform

Measures of the Effects of Improved Property Rights, a Stable Policy Environment, and Environmental Protection

Runsheng Yin

IN CHAPTER 2, D. Liu and Edmunds proposed that two features of China's reforms had significant, and sometimes opposing, effects on forestry in the first 15 years of reform:

- The transfers of some property rights from collectives to households improved the incentives for forest management and led to increases in forest cover and eventual increases in the standing forest volume. The land area in forest plantations in the market-driven nonstate sector increased at an annual rate of 6% from 1989 to 1993.
- A generally stable policy environment would reinforce the favorable effects of improved land tenure. However, the combined effects of a prior history of policy reversals, government foresters who were doubtful of the ability of households to manage forests, and periods of general inflation and subsequent reevaluation of some reforms created doubts about the long-term durability of the new household tenures. Where these doubts were sufficiently strong, households harvested their new forest stocks without replacement, and both forest area and forest volume declined.

This chapter treats these two features of China's reforms as propositions for rigorous econometric examination, using data from two regions: the Central/North/Northwest Farm Forest Region (CNNFFR), which experienced a surge in tree planting immediately after the introduction of reforms in 1978, and the Southern Collective Forest Region (SCFR), directly south of the CNNFFR. Forestry is a more important sector, government administrators exercised more influence over the sector and, therefore, forest policy and the actions of the gov-

Prior and partial versions of the material in this chapter appeared in Yin and Newman (1997) and Yin and Hyde (2000).

ernment forest bureaus played a greater role in household investment expectations in the SCFR than in the CNNFFR.

Households in the CNNFFR harvested the mature timber newly under their control, but they also replanted trees, and both the area and volume of their forest stocks grew. Annual timber harvests increased dramatically. In the more uncertain policy environment of the SCFR, households harvested without replacement as they were granted new management rights, and both the area and volume of the existing forests were drawn down. Forest cover more than doubled in the north, but it declined by more than 10% in the south. The econometric examinations in this chapter will confirm both propositions offered in Chapter 2. Indeed, the magnitudes of effect demonstrate that those propositions were central to forestry reform where it was more successful and they were a primary source of slower forestry-sector development and further deforestation where reform was less successful.

D. Liu and Edmunds in Chapter 2 and also J. Liu et al. in Chapter 3 raised the specter of burdensome timber taxes and regulations as constraints on forest investments. They suggested that when households added to the forest cover, it was for reasons other than the expected returns on marketable timber. Both chapters provided examples of households that planted other marketable forest crops—nontimber forest products such as fruit, rubber, and bamboo. In Chapter 1, Hyde et al. indicated the doubts that some have that individual households would invest in environmental improvements with communal benefits—because the households would not obtain sufficient benefits from those investments for themselves—regardless of their public merit. The argument is that public investments are necessary to obtain environmental benefits. Yet, in the examination of the CNNFFR in this chapter, it becomes apparent that that the private benefits of forest-based environmental improvement in China's northern plains were large and that individual farmers did plant trees to obtain these benefits—as well as the more marketable benefits obtained from nontimber forest products such as fruit and bamboo.

The northern plains are a part of the CNNFFR. They are China's greatest grain-producing region, and they are a region of severe wind erosion—which diminishes the region's agricultural productivity. Reforms in household tenure improved the region's agricultural production. They also increased the incentive for conservation investments in agroforestry activities such as windbreaks, which improved agricultural productivity further yet. The regressions in this chapter will demonstrate that improved tenure explained as much as 59% of a 7.5-fold increase in productivity. Trees planted in response to the improved incentives for conservation investments were responsible for as much as another 19% of the improvement in agricultural productivity, and they accounted for the largest share of this region's expanding forest cover.

Improved Tenure and Stability in the Policy Environment

Rural Reforms and Farm Forestry Behavior: The Regional Experience

As the household responsibility system (HRS) swept through China's agricultural sector, the idea of a similar program for forestry developed its own appeal. The idea found easy acceptance by provincial and local officials in the CNNFFR—referred to hereafter as the "northern production region" and including the provinces or metropolitan areas of Beijing, Hebei, Henan, Liaoning, Shaanxi, Shandong, and Tianjin and parts of Anhui—because forestry was a less important sector in this region.[1] It was a simple matter for forest policy reform to follow the pattern of policy reform in the agricultural sector in this region. A household forest tenure system called "land carrying trees" quickly followed. Trees on or near contracted agricultural land were assigned to agricultural households, as were bare lands that were suitable for planting trees. A few of the larger shelterbelt networks and commercial forest plantations that remained in collective ownership were also contracted to household management. Overall, the share of contracted forestland in the northern forest production region rose to 91% by 1984 (Yin 1994).

Furthermore, the government allowed farmers in this region the freedom to sell timber at market prices. Harvest permits were unnecessary, sales to state procurement companies were not required, and taxes were low—only 3–8% of sales revenues. In general, these arrangements were acceptable to farmers, and they have prevailed (Yong 1987, Zhong et al. 1991).

The experience of the SCFR—referred to hereafter as the "southern production region" and including the provinces of Fujian, Guangdong, Guangxi, Guizhon, Hainan, Hubei, Hunan, Jiangxi, and Zhejiang and parts of Anhui, Jiangsu, Sichuan, and Yunnan—was very different. The forest authorities in this region doubted the validity of a variation of HRS for forestry. They denied household rights to forestland on the presumption that household operations were too small to support long-term forest development activities. Two conflicts arose, one due to the denial of property rights and one due to boundary conflicts with entrenched state forest enterprises operating on collective lands. In general, only fragmented and less productive forests and woodlots were contracted to household management and, at the end of 1984, more than 70% of the SCFR's forests remained under collective control. Households in this region typically managed less than one hectare of forestland for their own use.

Eventually, popular objection overcame the government's reluctance, and the southern authorities slowly adopted the Contract Responsibility System for collective forests.[2] Contracted timberlands and plots of private forest were merged until households held contracts for more than 60% of all collective forest land by 1986. Contractual agreements extended up to 50 years. Output quotas became a popular contract criterion, and households were entitled to retain their production surpluses in excess of the quotas.

The government in the southern production region also opened the timber markets in 1985. Prices rose rapidly, encouraging further timber harvests and speculation. However, southern farmers did not expand their reforestation and forest management activities. They were not convinced of the government policies (Yin and Xu 1987, Wang et al. 1991). They had experienced four government policy reversals regarding the household ownership and use of trees during the previous 20 years, and they doubted the government would sustain its new policies.

Unfortunately, the farmers' fears were justified. The central government became concerned with inflation and reevaluated some of its reform policies.

Meanwhile, the local authorities became concerned about the liquidation of forests and the stagnation of forest management. They reversed several policies in 1986. First, they restored tough regulations on harvest volumes and returned timber markets to the control of state procurement companies. Farmers had to apply for harvest and transportation permits for those trees that remained under their contracts, and then they could sell only at government procurement prices that were set at approximately half the market prices. Subsequently, the local authorities began reconsolidating contracted timberlands. They also raised forest taxes and fees on those farmers who retained household forests.

Numerical Evidence

Tables 4-1 and 4-2 summarize the experiences of the two regions. Table 4-1 summarizes the price trends and household landholding experiences between 1977 and 1990. The price evidence is for two sample provinces in each region. Real timber prices rose more than fourfold in all cases, with larger absolute price increases in the northern production region. These price increases should have been an incentive for forest investment. The southern policy shift from procurement prices to retail prices in 1985 doubled that region's price incentive, but even retail prices in that region were controlled at a level well below market prices. Of course, the southern production region's subsequent return to procurement prices removed the additional retail price incentive and confirmed the existing policy uncertainty.

The contract share indicates the opportunity for household decisionmaking. The measure of this share in Table 4-1 is the share of both agricultural and forestland previously managed by collectives that was converted to household contracts. This measure is better than a measure of forestland alone because some farmers reallocated land between agricultural and tree crops.[3] Household shares rose more rapidly in the northern production region after forestry reforms beginning in 1980, and the transfer to households was 90% complete by 1984. Household shares rose less rapidly in the south, and the government even rescinded some previous household contracts. Thirty-five percent of the collective forests in the southern production region were still excluded from household management in 1990.

Differences in the rate and final level of conversion to household contracts distinguished the two regions, but the critical factor was uncertainty. The reluctance of the southern authorities to complete the transition and their rescission of some contracts, together with their decision to return to procurement prices, left southern farmers in a state of uncertainty with respect to future prospects for forest management.

What were the farmers' responses? Table 4-2 shows the levels of timber harvests and forest cover for two provinces in each region for two critical years: 1977 and 1987. The 1977 data provide a baseline from which to assess policy departures, and the 1987 data reflect a time period sufficient to absorb the readjustments in the south that occurred after authorities reversed the earlier policy liberalization.

Table 4-1. *Regional Comparisons of Increasing Timber Prices and Improving Tenure*

	Northern production region			Southern production region				
	Fuyiang	Suxian		Fuzhou		Yichun		
Year	Market price	Market price	Contract share	Procurement price	Retail price	Procurement price	Retail price	Contract share
1977	119.8	117.6	0.0	28.3	66.1	23.0	66.5	0.0
1978	125.4	120.7	0.0	25.0	66.6	25.3	67.7	0.0
1979	136.3	122.0	0.0	38.1	70.1	35.3	73.4	0.0
1980	165.9	152.0	32.9	46.3	83.3	50.5	87.8	14.4
1981	201.3	184.0	67.0	45.8	97.1	61.1	97.7	29.5
1982	241.0	222.0	78.8	56.5	104.3	64.0	108.1	37.7
1983	262.2	255.1	84.5	48.8	105.5	55.0	109.7	43.3
1984	265.0	297.0	91.0	55.3	109.0	57.9	116.4	52.4
1985	310.8	368.9	91.0	84.0	174.5	99.5	165.9	64.8
1986	351.5	377.3	91.0	93.0	207.5	118.0	226.2	64.8
1987	436.0	457.3	91.0	126.0	289.8	177.4	346.3	64.8
1988	630.1	667.8	91.0	171.4	444.1	173.1	502.2	64.8
1989	741.7	787.0	91.0	169.3	436.7	189.9	473.2	64.8
1990	639.0	698.0	91.0	170.4	350.3	164.8	386.4	64.8

Notes: Prices are in yuan and are nominal, averaged across species and grades. *Contract share* is the ratio of forestland allocated to households to total collective forestland.
Sources: Price data are from financial balance sheets of local timber companies and government reports. Contract share data are from the provincial forest services.

Table 4-2. *Regional Differences in Forest Resources*

Production region	*Province*	*Inventory measure*	*1977*	*1987*
Southern	Jiangxi	Area	5.47	5.90
		Volume	298.60	242.10
	Fujian	Area	4.51	4.53
		Volume	430.35	396.65
Northern	Shandong	Area	1.60	1.60
		Volume	24.26	47.94
	Henan	Area	1.42	1.57
		Volume	68.22	91.52

Notes: Area values are millions of hectares of closed forests; volume values are million cubic meters of both open and closed forests.

Source: Data are from forest survey reports of the provincial forest services.

In the northern production region, policy reform clearly led to increases in both short-term harvest activities and long-term reinvestments, and the latter led to increases in standing timber volumes. In the southern production region, uncertainty with regard to long-term policy prevailed at the household level. Reform initially led to larger harvests but it failed to stimulate investments in timber management. Subsequent policy reversals in the south encouraged additional harvesting of the remaining household lands and further discouraged investment. Standing forest inventories declined even more, and aggregate southern harvest levels began to fall as well—despite government investments in the remaining share of collective forests that still were not contracted to households and also in the south's larger component of state forests.

Anhui Province provides another comparison of these policy-induced effects on harvests and longer-term timber investments. Anhui lies on the boundary between the two regions. The northern third of Anhui is an important agricultural area; its southern two-thirds lies in the southern timber production region. Anhui is the province where HRS contracts for agriculture were first established, and Anhui's farmers were particularly alert to the advantages of household management. Timber harvests in both northern and southern Anhui increased from 1977 to 1980, but harvest levels in southern Anhui have declined ever since. Standing timber volumes in northern Anhui increased almost fourfold between 1977 and 1989, but they declined almost 10% in southern Anhui despite the concentration of government timber subsidies and government silvicultural investments in the south (Yong 1987).

Analytical Approach

The argument that the two regions responded differently to the China's policy reforms can be tested analytically by regressing short- and long-term regional forest production responses (timber harvests and measures of the forest inventory, respectively) on the two major policy instruments (household contract shares and the degree of price liberalization) and other factors, as appropriate. Positive relationships between the policies and both harvests and forest inventory indicate that the policies had the desired favorable effects on forestry. Positive relationships between policies and harvests but negative relationships between policies and forest inventory indicate the opposite and, therefore, the important negative effect of the remaining (statistically unspecified) policy difference, the unstable policy environment in the south.

More formally,

$$\ln Y_{it} = \alpha + \beta_1 P_{it-1} + \beta_2 CS_{it-1} + \beta_3 T + \sum_{i=2}^{5} \gamma_i D_i + \epsilon_{it} \qquad (4\text{-}1)$$

where the outcome Y is harvest volume in the regressions measuring short-run effects. Y represents the forest inventory, either standing timber volume or forest area in the regressions that measure longer-term effects; P is the timber price faced by farmers, deflated by the rural industrial producer price index; CS is the ratio of contracted forestland to total collective forestland; T is a trend variable that captures any sequential effects contained in the dependent variables, including technical change and input availability; and the D_i are cross-sectional dummy variables. The subscripts i and t reflect the cross-sectional and intertemporal differences in the observations, respectively. α is a constant, the βs and γs are parameters to be estimated, and ϵ_{it} is an independently distributed error term with zero mean and no intertemporal correlation. The second and third terms on the right-hand-side, last year's price and the share of land under contract, are related to this year's production in order to account for adaptive expectations and a lag in the production responses.

Data

The regional data sets cover the period from 1978 to 1989 for four prefectures in the northern production region (Fuyang, Kaifeng, Suxian, and Zhoukou) and five in the southern production region (Fuzhou, Ganzhou, Ji-an, Nanping, and Yichun). Therefore, the northern data set contains 48 observations, and the southern set contains 60. Local timber companies provided the data on harvest volumes, procurement prices, and market prices. Surveys conducted by the government forestry agency in each province in roughly five-year intervals provided the measures of forest cover and standing volume. Intermediate annual measures of these two variables were projected with the help of average growth rates for the regions. The provincial statistics services provided the rural industrial producer price index. Finally, the provincial forest agencies provided data on contracted and total collective forestland.

The 12-year time period of these data, the sample size, and the estimations necessary to create the forest inventory and land area data raise two technical questions. First, is the time period sufficient for the detection of long-run changes in forest production? In fact, although 12 years may be insufficient for identifying a new long-run equilibrium in forestry, it should be long enough to

Table 4-3. *Regional Forest Dynamics, 1978–1989*

Production region	Annual forest change (%)					
	Harvests		Standing volume		Area	
	Mean	SD	Mean	SD	Mean	SD
Northern	0.189	0.259	0.075	0.053	0.112	0.088
Southern	0.009	0.126	–0.010	0.006	0.008	0.009

Notes: SD, standard deviation. See text for description of data.

discern the sources of regional transitions. Estimates of the average annual growth rates in the dependent variables (shown in Table 4-3) justify this claim. In the northern production region, timber harvests, forest stocks, and forest area grew at annual rates of 18.9, 7.5, and 11.2%, respectively. In the southern production region, in contrast, the same measures grew very slowly or even declined. (The north–south differences in volume and area are statistically significant.) These differences seem to indicate substantially different regional production responses to the reform policies.

Second, the different prefecture land areas and initial forest volumes create an econometric problem that can be addressed by normalizing each dependent variable within a prefecture by its 1980 level. The cross-sectional dummy variables control for unidentified local differences in factors such as weather and site quality.

Estimated Results

Tables 4-4 and 4-5 show the regression results for the northern and the southern production regions, respectively. In general, the R^2 values indicate good equation fits in both regions. Dropping the trend variables only reduces the R^2 values by a small amount—probably because the trend variables and the two policy variables moved together during the period of this analysis. Removing the trend variable increases the policy parameters (in all cases) for the north but decreases them (in two of three cases) for the south. Therefore, the trend variables prevent any overstatement of policy effects in the north or understatement of policy effects in the south.[4]

The coefficient estimates confirm all expectations. In the north, harvests (columns 1 and 2) responded positively to both increasing prices and improved tenure. In the south, harvests responded positively to improved tenure but displayed no significant relationship to increasing prices. The prospect of losing control of the resource dominated any prospect of increasing future prices in the south. The apparent conclusion is that southern farmers with uncertain tenure harvested and sold what they could while they controlled the resource, regardless of future price prospects.

In the north, standing forest volume (columns 3 and 4) and forest area (columns 5 and 6) responded positively and significantly to both policy variables. In the south, the relationships between these measures of forest investment and the two policy variables are small, insignificant, and generally negative when the

Table 4-4. *Supply Response Regressions for the Northern Forest Production Region*

	Harvests		*Standing volume*		*Area*	
	(1)	*(2)*	*(3)*	*(4)*	*(5)*	*(6)*
Constant	−1.11***	−1.21***	−0.37***	−0.41***	−0.59***	−0.64***
	(3.82)	(3.77)	(4.99)	(4.32)	(4.83)	(4.53)
Price, P_{t-1}	0.29*	0.41***	0.13***	0.18***	0.17***	0.24***
	(1.93)	(2.36)	(2.98)	(3.39)	(2.51)	(3.06)
Tenure, CS_{t-1}	0.003	0.009***	0.002***	0.005***	0.004***	0.007***
	(0.96)	(4.11)	(3.07)	(7.40)	(3.29)	(7.32)
Trend, T	0.10***		0.04***		0.05***	
	(3.19)		(5.14)		(3.81)	
Prefecture dummies						
D_1	0.03	0.01	−0.11***	−0.12***	−0.04	−0.06
	(0.03)	(0.01)	(3.44)	(3.01)	(0.77)	(0.94)
D_2	1.01***	0.95***	0.11***	0.09**	0.26***	0.23***
	(8.55)	(7.31)	(3.65)	(2.20)	(5.20)	(3.98)
D_3	0.52***	0.47***	0.13***	0.11***	0.28***	0.26***
	(4.37)	(3.60)	(4.26)	(2.84)	(5.62)	(4.46)
Degrees of freedom	41	42	41	42	41	42
R^2	0.87	0.84	0.95	0.91	0.93	0.90

Note: Asymptotic *t*-tests shown in parentheses.
*Significant at the 0.10 level. **Significant at the 0.05 level. ***Significant at the 0.01 level.

trend variable is included in the regression.[5] This is the expected result if farmers harvest at their first opportunity and disinvest (harvest any subsequent growth without reinvestment) whenever permitted the renewed opportunity to harvest.

These regional differences support the proposition that policy uncertainty works against long-term forest investment. That is, in the northern production region, longer-term investments in forestry followed improved land tenure and increasing prices acted as an additional investment incentive once land tenure was more secure. In the southern production region, the same two desirable policies failed to encourage forest investment because farmers had no confidence that the policies would endure long enough for them to obtain the benefits of their investments.

Finally, the trend variable is always positive in the north, whereas it declines in the harvest and investment regressions for the south. This difference is consistent with the expected downward trend in southern investments due to the insecurity of household land tenure. With declining investments in the south, the standing forest volume must also decline, implying that future harvests must decline as well. The land area in forests should follow the same pattern as the standing forest

Table 4-5. *Supply Response Regressions for the Southern Forest Production Region*

	Harvests		*Standing volume*		*Area*	
	(1)	*(2)*	*(3)*	*(4)*	*(5)*	*(6)*
Constant	0.02	0.03	0.04***	–0.04***	–0.03***	–0.03***
	(0.37)	(0.42)	(4.23)	(4.27)	(4.05)	(4.04)
Price, P_{t-1}	–0.09	–0.23*	0.01	–0.03*	–0.004	0.04***
	(0.62)	(1.84)	(0.32)	(1.78)	(0.18)	(2.53)
Tenure, CS_{t-1}	0.006**	0.002	–0.001	–0.001***	–0.001	0.001***
	(2.27)	(1.38)	(0.01)	(4.67)	(1.08)	(3.24)
Trend, T	–0.05*		–0.01**		0.01***	
	(1.81)		(2.09)		(2.67)	
Prefecture dummies						
D_1	–0.06	–0.05	–0.04***	–0.01***	0.01	0.01
	(1.05)	(1.13)	(5.44)	(5.45)	(1.05)	(1.24)
D_2	0.07	0.06	0.01	0.01	–0.01	–0.01
	(1.24)	(0.98)	(1.37)	(0.96)	(1.22)	(0.69)
D_3	0.05	0.03	–0.04***	–0.05***	0.04***	0.04***
	(0.88)	(0.48)	(4.45)	(2.84)	(4.65)	(5.47)
D_4	0.35***	0.34***	–0.03***	–0.03***	0.02***	0.02***
	(5.91)	(5.68)	(3.48)	(3.48)	(2.96)	(2.96)
Degrees of freedom	52	53	52	53	52	53
R^2	0.55	0.52	0.81	0.79	0.77	0.74

Note: Asymptotic *t*-tests shown in parentheses.
*Significant at the 0.10 level. **Significant at the 0.05 level. ***Significant at the 0.01 level.

volume. That it does not is a reflection of the larger size of government reforestation programs in this region. Unfortunately, an increase in the forest area at the same time the standing forest volume is decreasing means that the average quality of the resource is also decreasing. Average stocking in the south declined to about 43 m^3/ha in the late 1980s, or less than half of the world average.

The estimated regression coefficients can be used to calculate the contributions of each reform policy to economic growth, and these results can be compared with measures of the contribution of similar agricultural policies. The comparison with agricultural policy provides insight to the relative importance of the forest policy reforms as well as additional insight to the general reasonableness of the findings. The procedure is to multiply the estimated coefficients by 100 (because the coefficients are in a semilog form), calculate the average change in each variable as the weighted difference in magnitude of that variable among all cross-sections between the first and last sample periods, and multiply the results from the

Table 4-6. Accounting for Forest Growth, 1978–1989

Region and explanatory variable	*Change in variable, 1978–1989*	*Harvest*		*Standing volume*		*Area*	
	(1)	*(2)*	*(3)*	*(4)*	*(5)*	*(6)*	*(7)*
Northern production region							
Price	1.02	28.57	29.14	12.59	25.12	17.17	17.51
			(15.84)		(26.51)		(15.11)
Tenure	89.35	0.28	25.29	0.23	20.64	0.41	36.45
			(13.75)		(21.78)		(31.46)
Trend	11.00	9.67	106.37	3.98	43.78	4.89	53.79
			(57.82)		(46.21)		(46.43)
Residual			23.18		5.21		10.51
			(12.59)		(5.50)		(7.00)
Total growth			183.98		94.75		115.86
			(100.00)		(100.00)		(100.00)
Southern production region							
Price	1.66	–8.91	–14.79	0.81	1.34	–0.38	–0.63
			(–142.70)		(13.90)		(–7.60)
Tenure	63.72	0.63	40.14	–0.09	–5.73	–0.04	–2.55
			(387.50)		(–59.44)		(–30.79)
Trend	11.00	–4.54	–49.94	–1.05	–11.55	1.07	11.77
			(–482.00)		(119.80)		(141.98)
Residual			14.23		6.30		0.30
			(137.40)		(65.35)		(3.60)
Total growth			–10.36		–9.64		8.29
			(100.00)		(100.00)		(100.00)

Notes: Column 1 records the change in the explanatory variable. Columns 2, 4, and 6 are the estimated coefficients (×100). Columns 3, 5, and 7 are the contributions to growth and are the products of column 1 × column 2, 4, and 6, respectively. Numbers in parentheses are the percentage share contributions to total growth.

first two steps for each productive factor. The final step is to compute the contributions of the residuals by dividing total growth by the degree of fit (R^2).

Table 4-6 illustrates the calculation and its results. In the northern production region, price changes and tenure reform shared approximately equally in 30% of the 184% growth in harvests between 1978 and 1989. Prices and tenure policy also explain approximately half of the increase in long-term forest investment in this region (48% of the 95% growth in standing volume and 47% of the 116% expansion in forest cover). Comparable estimates for the agricultural sector show that rising prices and improved tenure accounted for 57% of the 42% growth in agricultural output from 1978 to 1984; 15 points of the 57% was due to increasing agricultural prices, whereas the remaining 42 points were due to tenure reform (McMillan et al. 1989, Lin 1992, Wen 1993). Thus, the policy reforms stimulated a relatively greater expansion in forestry than in agriculture. This is an important but unsurprising observation because secure tenure becomes more critical the longer the term of investment, and forestry is generally a longer-term activity than agriculture.[6]

Table 4-6 also shows that the increase in harvests in the southern production region was due almost exclusively to improved tenure. This observation is consistent with the reasoning that tenure reform dominates price increases in the presence of uncertainty. Managers must be confident of their control of the resource before price increases will induce them to invest. Table 4-6 also shows that, in contrast, increases in the forest cover in the south were due to something other than improved tenure. They were probably due to government silvicultural investments on state forestlands—as suggested earlier in this section.

More Evidence

Additional confirmation of the different incentive structures in the two regions can be obtained by decomposing their timber prices and reviewing the conditions of their physical forest resources for the years after the initial forestry reforms. Household surveys of 100 farmers, in each of the original sampling areas, conducted after completion of the econometric analysis, revealed that most farmers in the northern production region believed that the new tenurial arrangements and the open market for timber transactions were satisfactory and should be maintained. The majority of farmers in the southern production region believed that their land tenure was still uncertain, and they indicated that this insecurity contributed to their reluctance to plant and manage trees. In addition, approximately 80% of the southern farmers felt that the controlled procurement prices still facing them in the mid-1990s were low and that the burden of taxes and fees imposed on them was heavy.

Farmers in the north were not required to sell their timber to state procurement companies. They obtained market prices for their timber and kept more than 90% of total revenues from their timber sales in 1990. In contrast, individual farmers and collectives in two representative southern prefectures (Sanming and Fujian) kept only 31% of total revenues from their timber sales. The remaining 69% was divided among taxes (23%), forest bureau fees (26%), local government charges (7%), and timber procurement company transactions costs (13%). These burdens on southern timber sales decreased the farmers' incentives for forest investment and supported their perception that timber prices were low.

One might inquire whether physical differences between the two regions affected production. The substantial increase in forested area in the northern production region and the slight increment in the southern production region might indicate that more land was available for reforestation in the north. In fact, this was not the case. The percentage of land with forests and trees in the north in the mid-1990s was more than 100% of the land regarded as available for forestry—because northern forests were well stocked and various agroforestry regimes placed trees on some lands that were primarily agricultural. In the south, however, forested land accounted for only 55% of the total area regarded as available for forestry. It would have been impractical to introduce much additional management on the well-stocked northern stands. However, the approximately 70% of southern stands that were understocked or contained only young trees indicates considerable opportunity for expansion in that region (MOF 1988).

Environmental Protection as an Investment Objective in the Northern Plains

If taxes, controlled prices, and restrictive permits on timber harvests and shipments diminished the incentives for forest management, and policy uncertainty diminished those incentives further yet, then what explained the expanding forest cover on the collective forestlands of most regions? D. Liu and Edmunds observed in Chapter 2 that households planted for forest products that were not subject to the high timber taxes and timber shipment restrictions. J. Liu et al. made a similar observation in Chapter 3.

This section of the chapter considers whether tree planting to provide various environmental services could have been another source of the increase in forest cover—particularly in severely deforested regions where the environmental gains may have been large and rapid. An example from China's northern plains demonstrates that this proposition is reasonable—despite recognition that forest-based environmental services are public goods and that their benefits do not all accrue to the household that plants and manages them.

The northern plains are roughly comparable to the northern production region discussed in the previous section. They are the eastern part of the CNN-FFR, including the provinces or metropolitan areas of Beijing, Hebei, Henan, Liaoning, Shandong, and Tianjin; the northern parts of Anhui and Jiangsu; and the central part of Shaanxi. The northern plains were not subject to burdensome timber taxes and other restrictions on timber harvests and management. But these facts do not affect our claim. If households in this region obtained sufficient private advantage to justify planting trees, then households in other regions also may have perceived the private environmental advantages of increasing their forest cover. The econometric assessment in this section demonstrates, once more, that the receipt of improved tenure was a critical precondition to investment and that, in this case, the household gains from private investment in environmental improvement were large indeed.

The Agroforestry Experience

Although China has a rich tradition in agroforestry extending over hundreds of years, its current experience is a matter of the last 20 years (Zhu et al. 1991). In the northern plains, few trees—let alone forests—survived the wreckage of war and exploitation that preceded the founding of the People's Republic in 1949. Average agricultural grain production was approximately 800 kg/ha, and many hectares of farmland were abandoned due to desertification, salinization, changes in the course of the Yellow River, and other environmental stresses (*Henan Yearbook* 1986, *Shandong Yearbook* 1988). Living conditions were abysmal. Even in the 1990s, approximately 243 million people lived on a land base of 464,000 km^2, of which only 26 million ha was arable farmland—or about 0.08 ha per capita (Zhong et al. 1991). Furthermore, the base of arable land was declining at an annual rate of 0.8% while the region's population was growing at a rate of 1.5% annually (*China Agricultural Yearbook* 1985–1989).

The central government began directing attention to land rehabilitation in the early 1950s. Its initial focus was on various environmental engineering and biological activities, including afforestation projects for windbreaks and for controlling desertification and salinization. The limited new forest resources were rapidly depleted, however, after the collectivization and industrialization movements that began in 1958. Collectivization raised doubts about the ownership security of forest resources, and backyard furnaces for steel production consumed most of the remaining wood for fuel (*China Forestry Yearbook* 1986).

The economic adjustments of the early 1960s promoted multipurpose afforestation activities such as shelterbelts and intercropping for the purposes of protecting the environment and improving agricultural productivity. Unfortunately, the Cultural Revolution and the political struggles that dominated Chinese society for the next decade restricted any real opportunity for environmental conservation or economic growth. Agricultural and forest production stagnated. Per capita annual income in the communes was only 62.8 yuan in 1976 (about US$25); more than one-third of rural households were in debt; and many households lived with insufficient fuel, clothing, and housing. About 100 million people suffered from food shortages (Lu 1986).

After the Cultural Revolution, economic development became the policy priority and market mechanisms for allocating resources began to replace quotas and price controls. In addition, the contract-based HRS replaced collective production teams as the basic organization for rural production. Agricultural production has expanded at an unprecedented pace ever since. Grain yields in the northern plains surpassed 6,000 kg/ha/year (*China Agricultural Yearbook* 1988) by the late 1980s, and this region is now among the most productive agricultural regions in the world.

China's agriculture has witnessed phenomenal technological adjustment over the past several decades, including the expansion of its irrigation systems, the application of chemical fertilizers and pesticides, the introduction of improved seeds, and the use of plastic sheeting to control weeds and concentrate heat. These technologies have their limitations, however. Several problems have arisen, including a decline in the upper layers of groundwater, a decrease in soil organic content, an increase in the vulnerability of crops to pests and diseases, and extensive crop susceptibility to desiccating early summer winds. Maintaining a sound environment conducive to steady increases in productivity appears to be beyond the reach of modern technology alone.

Properly distributed forest cover could be an important complementary technology for sustaining agricultural production. Trees reduce wind velocity and modify solar radiation, thereby regulating air and soil temperature and increasing

Table 4-7. *Forest Resources in Select Provinces of Northern China, 1988*

Province	Area: Shelterbelts (× 1,000 ha)	Area: Intercropping (× 1,000 ha)	Area: Woodlots (× 1,000 ha)	Area: Four sides (billions of trees)	Standing timber volume (m^3)	Percentage of total land area
Henan	2,200	1,987	223.3	1.13	34.24	12.1
Shandong	3,800	1,335	400.0	1.28	35.05	7.4
Jiangsu	2,000	300	482.0	1.20	25.00	8.2
Anhui	1,434	40	276.7	1.15	19.14	13.9
Hebei	3,286	47	208.7	0.88	29.00	9.7
Shanxi	976	700	108.0	0.04	17.00	15.0
Shaanxi	967	90	300.0	0.05	7.96	7.2
Total	14,663	4,499	1,698.7	5.73	167.39	

Notes: Shelterbelt is another term for "windbreak." *Intercropping* refers to agricultural and tree crops grown intermixed in the same field; the area reported is the intercropped area under tree cover. *Woodlots* refer to small forested areas used for fuelwood, construction wood, and perhaps other forest products. *Four sides* refers to trees planted around villages and farmhouses, and along roads and waterways. *Standing timber volume* excludes the extensive area and volume of natural forests; this volume is outside any usual conception of agroforestry.

Source: Afforestation Bureau (MOF 1988).

field moisture. Trees also improve soil nutrients and contain erosion. Trees do compete with agricultural crops for light and water but the competition is not great at the low levels of tree cover that still prevail in this region. China's northern plains were largely deforested by 1970, and the overall potential for tree cover to contribute to agricultural production was probably positive.

In fact, China's agricultural regions in general, and the northern plains in particular, experienced an upsurge in tree planting after the introduction of rural reforms in 1978. Forest cover in the northern plains increased from about 5% in 1977 to 11% in 1988, thereby greatly easing local shortages of fuelwood and small construction timber and substantially improving the environment for agriculture (Yong 1989, Zhong et al. 1991). Table 4-7 shows that by 1988, the region possessed 14.66 million ha of forest shelterbelts, 4.5 million ha of farmland intercropped with trees, 1.7 million ha of small woodlots, and more than 5.7 billion trees planted around the "four sides" of villages, farm houses, roads, and waterways (MOF 1988).

Analytical Approach

The effects of these new trees on the agricultural environment can be measured with an agricultural production function. The expectation is that agricultural productivity is a function of standard agricultural inputs: land, labor, and capital of various sorts. The important policy reforms and the level of agroforestry activity are production shifters. Policies such as improved tenure provide an incentive to use all inputs more efficiently, and production increases. Improved tenure also prompts some farmers to plant trees as a means of improving their agricultural environment. The improved environment affects the overall conditions in which agricultural activity operates and thereby improves agricultural productivity. If the estimated parameters on the production shifters are positive and significant, then it is reasonable to infer that improved tenure provides the anticipated incentive for improved productivity and, as it does, it permits a longer-term management perspective. Farmers respond by planting trees to provide a series of environmental services that complement agricultural productivity.

The basic function takes the form

$$\begin{aligned}\ln Y_{it} &= \alpha + \beta_1 \ln A_{it} + \beta_2 \ln K_{it} + \beta_3 \ln L_{it} + \beta_4 \ln F_{it} \\ &\quad + \gamma_1 HRS_{it} + \gamma_2 CM_{(it-1)} + \gamma_3 ES_{it} + \gamma_4 T + \sum_{i=2}^{5} \delta_i D_i + \epsilon_{it} \qquad (4\text{-}2)\end{aligned}$$

where Y is aggregate production, and the terms in logs are the usual inputs to agricultural production: land (A), labor (L), durable capital (K), and less durable capital inputs such as fertilizer (F). The subscripts i and t refer to regional distinctions (prefectures) within the sample region and the data year, respectively; α is a constant; the βs, γs, and δs are parameters to be estimated; and ϵ_{it} is an independently distributed error term with zero mean.

The second line of the equation describes the production shifters. *HRS* is the proportion of all production teams converted to the household responsibility system (the measure of tenure reform), *CM* is the ratio of crop to manufactured input prices (an indication of the degree of market reform that gradually made agriculture a relatively more attractive economic activity), *ES* is a measure of the environmental services provided by agroforestry activities, T is a time trend, and the D_i are cross-sectional dummy variables.[7] Two institutional reforms had tremendous effects on China's agricultural performance: the tenurial shift from collectives to the HRS and market reform. The ratio of crop to manufactured input prices (*CM*) is an indication of the degree of market reform that gradually made agriculture a relatively more attractive economic activity. [Lin's (1992) classic analysis of Chinese agriculture used the same ratio to explain market reform (also lagged one year to account for adaptive expectations), and a similar model of agricultural production for the same period.] *HRS*, *CM*, and *ES* affect the entire production function because they improve the entire agricultural environment. They improve the conditions for all other agricultural inputs: labor, equipment, seed, and fertilizer. This formulation is equivalent to identifying agroforestry as a technological innovation from the perspective of land management (Hayami and Ruttan 1985). Finally, the cross-sectional dummy variables (D_i) isolate the effects of local variations in factors such as soil quality and the extent of irrigation, and the time trend T captures the effects of other unidentified periodic production shifters.

The Cobb–Douglas functional form is well-suited for these data. China's generally unreliable input price data prevent the use of a pure cost function, and the limited number of available observations restrict the use of a more flexible form for the function. These limitations are not a problem, however. Robust Cobb–Douglas results should be sufficient to test the HRS and agroforestry propositions.

Data

The measure of aggregate agricultural output is the sum of the annual physical products of grain, cotton, and oilseed (which account for more than 90% of production in the region) multiplied by their official prices in 1980. This means that the 1980 price is the common denominator for aggregation. Official agricultural prices were the only received prices in 1977, and official procurement prices still did not differ greatly from market prices by 1990, the final period for the analysis. [Lin (1992) also used an official price to obtain a measure of aggregate output.]

The measures of agricultural inputs are total sown area *A*, total farm labor force *L*, agricultural capital *K*, and chemical fertilizer *F.* Sown area is a better measure than cultivated area because sown area captures the effect of multiple cropping. The farm labor force is the total rural labor force minus those laborers engaged in nonagricultural sectors. The measure of agricultural capital is the total horsepower (hp) of farm machinery, plus draft animals (valued at 0.7 hp per animal). This measure may overestimate total agricultural capital because many farm machines are used in other activities. Therefore, an alternate specification can be considered—with draft animals alone as the measure of capital. The measure of chemical fertilizer is simply the gross weight of fertilizer consumed. Irrigation is another important capital input, but independent irrigation data are unavailable. Irrigation machinery, however, is part of the farm machinery that is captured in the measure of capital. Therefore, the coefficient for capital also reflects the impact of irrigation.

Finally, the data originate from five representative prefectures in Shandong Province for the 13-year period from 1977 to 1990—which means that the data set contains 65 observations. The period from 1977 to 1990 is long enough to incorporate most of the effects of the changes in agricultural reforms that began in 1978. Shandong, the largest and most diverse province in the northern plains, is also geographically central in that region. Shandong's Forest Bureau provided the data for intercropped and shelterbelt areas from the forest surveys of 1977, 1982, and 1988. The forest inventory data for intermediate years are estimates based on forest survey data and forest growth rates. The provincial Statistics Bureau assisted in compilation of the agricultural data. Table 4-8 summarizes (for the entire province) the full data set in index form, where 1980 is the index year.

Reflections on the Functions of Agroforestry

Tree cover (in hectares) is a reasonable proxy for the environmental services provided by trees (represented by *ES* in Equation 4-2), as long as the trees are well established and well distributed in the immediate areas of agricultural production. This proxy and the analysis might be refined further by introducing separate agroforestry production shifters for trees intercropped with agricultural products

Table 4-8. *Indices of All Variables for Shandong Province, 1978–1990*

Year	*Output*	*Land*	*Labor*	*Capital*	*Fertilizer*	*HRS*	*Market reform (CM)*	*Agroforestry*
1978	87.18	101.34	91.88	84.02	71.58	0.00	79.62	17.68
1979	90.12	100.60	96.47	93.46	81.56	0.01	82.05	19.74
1980	100.00	100.00	100.00	100.00	100.00	0.12	100.00	22.78
1981	120.82	97.89	102.80	109.87	119.73	0.40	99.81	25.75
1982	148.31	97.39	103.32	121.73	142.72	0.67	105.61	28.96
1983	175.57	99.84	103.96	145.43	167.60	0.83	104.91	32.25
1984	199.92	102.84	105.08	158.53	171.10	0.95	105.21	35.01
1985	179.99	105.16	104.29	177.81	179.30	0.99	102.29	38.27
1986	180.72	107.25	104.40	196.65	177.04	0.99	96.36	42.10
1987	195.76	106.32	105.38	216.05	176.92	0.99	103.84	46.14
1988	182.83	106.60	107.61	240.89	194.30	0.99	106.99	46.88
1989	192.31	105.72	109.44	245.22	201.98	0.99	114.50	47.29
1990	195.33	99.15	104.16	248.75	215.55	0.99	153.54	47.06

Notes: All variables are defined in text. *Agroforestry,* the index for combined shelterbelt and intercropped land shares, excludes smaller areas in woodlots and "four sides" areas, which were not planted for their impacts on agricultural productivity.

Source: Shandong Yearbook 1985 through 1992.

as well as for trees in shelterbelts. The new proxies are the ratios of area in the respective agroforestry activities to total farm area.

Trees contribute to the agricultural environment through five physical mechanisms:

- reducing wind velocity and wind erosion,
- controlling sheet and rill erosion,
- modulating solar radiation and regulating soil and air temperature,
- increasing field moisture, and
- adding soil nutrients.

Desiccating winds have been a massive problem in other regions of China, and shelterbelts are specifically designed to control them. (The Green Great Wall is the world's largest shelterbelt project.) Farmers in the northern plains are particularly alert to their potential. Intercropped trees are not planted densely enough to form good wind barriers, and their effect on wind erosion is minimal, but they contribute to the agricultural environment through the other four mechanisms. Intercropping is particularly important in the northern plains for its contributions to soil temperature, field moisture, and soil organic matter.

Woodlots and "four sides" trees have altogether different objectives, and the combined area they cover is much smaller than the area covered by shelterbelts and intercropped trees (Table 4-7). Woodlots (smaller blocked forest stands) are primarily sources of construction timber and fuelwood, although they may help control wind erosion and some woodlots protect riverbanks and also help with flood control. "Four sides" trees are planted to improve the ambient environment in areas of heavy human activity, such as the vicinity of farmhouses. These objectives are not so clearly complementary with agriculture. Indeed, they reduce the land available for agricultural crops. The local forestry agencies recognize the different forest objectives and compile data on woodlots and "four sides" trees in a different form from their data on shelterbelts and intercropping. All of these are reasons to exclude trees on these latter two categories of land use from the analysis.

One other consideration is important: the lag between the initial establishment of tree seedlings and the effect of forest cover on the environment. This lag may be short for intercropped trees and shelterbelts because the environment was so severely depleted and because farmers in the northern plains plant fast-growing trees. However, the positive effects, especially for shelterbelts, continue accumulating over long periods as their wind-reducing effect only reaches its maximum with mature trees.[8] Therefore, the impacts of young trees accumulate rapidly, but the full impacts of shelterbelts continue increasing well beyond the 13-year period of these data. A one-year-lag showed the best statistical fit for the period between the initial observation of trees on the land and their first effects on agricultural productivity. This lag validates the assumption that agricultural reform (and improved tenure in particular) preceded agroforestry activities—by at least one year.

Empirical Results

Table 4-9 shows the regression results. They were corrected for heteroscedasticity due to different prefecture sizes by dividing all input and output variables by the area of cultivated land in the prefecture in 1980. Otherwise, the two regressions are fixed-effects models (with a time trend). (Autocorrelation was detected in the error terms, but correcting for it produced little improvement.) The first two columns of results report the aggregate production function with the two different definitions of capital. The third column eliminates the physical inputs and focuses on the production shifters as a means of confirming the agroforestry findings from the production functions. In general, the statistical fit indicated by the R^2 is satisfying for all three regressions, the coefficients on the agroforestry variables are positive, and the coefficient on intercropping is always significant. The coefficients on most of the remaining variables satisfy expectations, and many of them are statistically significant.

The coefficient on tenure reform is large and significant, whereas the coefficient on market reform is small and insignificant. These results are consistent with the findings of McMillan et al. (1989) and Wen (1993) for China in general. They provide specific confirmation for Shandong and the northern plains that conversion to the HRS was the primary source of growth in agricultural productivity. The much smaller effect of market reform is not surprising. The data in Table 4-8 show that the changes in this index were too small in most years to build a case for a significant contribution to agricultural revenues.

Table 4-9. *Regression Results for Agricultural Production*

Variable	*Specification with full capital*	*Specification with draft animals as the only capital*	*Focus on production shift variables*
Intercept	0.361	2.618*	2.456*
	(0.660)	(3.829)	(13.333)
Agroforestry			
Shelterbelt	0.005	0.001	0.010
	(0.767)	(0.206)	(1.204)
Intercropping	0.007*	0.006*	0.007*
	(2.519)	(2.041)	(2.033)
Tenure reform (HRS)	0.361*	0.551*	0.663*
	(4.225)	(6.368)	(7.909)
Market reform (CM)	0.0002	–0.0001	0.0001
	(0.336)	(–0.277)	(0.083)
Capital	0.536*	0.346*	
	(4.588)	(5.642)	
Fertilizer	0.435*	0.311*	
	(4.212)	(2.966)	
Land	–0.382	–0.043	
	(–1.502)	(–0.195)	
Labor	–0.064	–0.172	
	(–0.189)	(–0.549)	
Time trend	–0.065*	–0.035*	
	(–4.525)	(–3.405)	
Prefecture dummies			
D_1	–0.250*	–0.510*	–0.467*
	(–2.983)	(–6.223)	(–7.595)
D_2	–0.046	–0.290	–0.048
	(–0.277)	(–1.884)	(–0.397)
D_3	0.214*	–0.082	–0.172
	(2.110)	(–0.979)	(–1.624)
D_4	–0.118	–0.508*	–0.625*
	(–1.048)	(–4.724)	(–5.279)
Degrees of freedom	51	51	55
adj. R^2	0.958	0.964	0.914

Notes: Values in parentheses are asymptotic *t*-statistics.
*Significant at the 0.01 level.

The coefficients on the two measures of capital are not significantly different from each other, and the choice between them does not alter the agroforestry results or the overall statistical reliability of the regressions. This finding is consistent with the observation that farmers with more machinery also tend to possess more draft animals, an observation that would be reasonable if draft animals were the major source of field traction and if equipment such as small tractors was used for other purposes. Finally, the coefficient on fertilizer is also positive and significant.

The coefficients on land and labor are negative but insignificant. Table 4-8 reveals relatively little variation in the data for the land variable. In fact, land area declined slightly while output expanded rapidly. Therefore, it is not surprising that land is statistically uncorrelated with output.

Labor is a different question. The northern plains population has been growing at an annual rate of 1.5%, even while the agricultural land base has been declining. Table 4-8 shows that the farm labor force increased in Shandong, even while it fell by approximately one-half in China as a whole. Therefore, our estimated labor coefficient is not surprising. It is even less surprising when we consider that China's official agricultural labor statistics are calculated as residuals after all other rural employment has been identified for the region, rather than as actual estimates of agricultural labor. Residuals are often poor predictors.[9]

The sum of the coefficients on productive factors is less than one. This indicates that the production process is characterized by decreasing returns to scale—which implies that productivity increases have become more and more difficult to obtain over time. This finding should raise concerns about the outlook for regional agriculture and heighten the importance of exogenous factors such as policy shifts (e.g., *HRS* and *CM*) and environmental improvements (e.g., agroforestry) that can improve productivity. Because the northern plains is a major source of China's national production and a leader in China's recent overall expansion in agricultural productivity, this finding also raises questions about the sustainability of national agricultural development without further improvements in policy reforms as well as further improvements in the agricultural environment.

Finally, the time trend is negative and significant. This finding differs from Lin's (1992) finding of a positive trend for all of Chinese agriculture. Our different observation may indicate only that the most important regional technological changes were embedded as quality improvements in capital and fertilizer. [This would be consistent with Stone's (1988) discussion for all of China.] It also could be due to the sharp reduction in government investment in the northern plains' agricultural infrastructure, including poor maintenance and inadequate expansion of irrigation and other public production facilities and limited access to extension services (Feder et al. 1992). The irrigation infrastructure is greater in the northern plains than in other regions of China, and the decline in the budget allocated to it was notable.[10]

The third specification adds conviction to the findings of the first two regressions. It eliminates the usual agricultural inputs and the trend variable in order to focus on the statistical robustness of the production shifters. Tenure continues to be the dominant policy variable. The intercropping coefficient is constant at 0.006–0.007 and significant, regardless of the data or model specification. The shelterbelt coefficient is positive but (insignificant and) probably smaller, regardless of the data or model specification.[11] These results suggest that the capacity of intercropped trees to regulate the temperature and moisture of soil and air and improve soil nutrients was more important to Shandong farmers than the more limited capacity of young shelterbelts to control wind velocity and wind erosion. This may change. As the trees mature beyond the 13-year period of our data, the wind-controlling effects of shelterbelts will increase, and any conclusion about the relative importance of shelterbelts and intercropping may change. The government has anticipated the growing potential of shelterbelts. Government experiment stations have explicitly worked with one species, Paulownia (*Paulownia spp.)*, for shelterbelts and Paulownia shelterbelts in Shandong increased from 113,000 ha in 1977 to 775,600 ha in 1988, when they made up 20% of the total shelterbelt area in China (*Shandong Yearbook* 1988).

Two immediate conclusions can be drawn from these observations:

- The policy change from collective management to the HRS was the dominant source of improvement in agricultural productivity through 1990. Farmers planted trees with confidence that they would recover the benefits of their investments.
- Agroforestry in general, and intercropping in particular, had a rapid and positive impact on agricultural production in Shandong and probably throughout the northern plains.

Factor Shares

The contributions of HRS and agroforestry can be examined further by decomposing the growth in agricultural output into its factor shares. The procedure is comparable to the procedure for Table 4-6. Table 4-8 shows that the tenure reform associated with HRS was essentially complete by 1984. Therefore, it is reasonable to separate the decomposition of effects into two periods, 1977–1984 and 1985–1990. Table 4-10 presents the decomposition results for the second regression in Table 4-9. Columns 4 and 6 of Table 4-10 report the factor shares. The first number in these columns is the absolute share of agricultural growth, and the parenthetical expression underneath is the percentage share.

The conversion to HRS explains more than one-half of the total growth in the first period. This observation is roughly comparable to Lin's (1992) estimate of 47% of agricultural growth nationwide due to HRS. McMillan et al. (1989) and Fan (1991) made similar observations. Aside from the change in tenure, the capital, fertilizer, and environmental services arising from intercropping were the major contributors to productivity growth. Agroforestry—including the combined impacts of shelterbelts and intercropping—contributed moderately, increasing in absolute share over the two periods (from 4.97 in 1977–1984 to 5.39 in 1985–1990). Its proportional share was only 5.42% during the earlier period, when HRS had its initial great impact and most agroforestry trees were

Table 4-10. *Accounting for Productivity Growth*

		1977–1984		*1985–1990*	
	Regression coefficient	*Average change in variable*	*Contribution to growth*[a]	*Average change in variable*	*Contribution to growth*[a]
Inputs					
Capital	0.346	74.51	25.78 (28.05)	86.69	29.99 (107.66)
Fertilizer	0.311	99.52	30.95 (33.67)	30.88	9.60 (34.47)
Land	–0.043	1.50	–0.065 (–0.07)	2.88	–0.12 (–0.44)
Labor	–0.172	13.20	–2.27 (–2.47)	4.36	–0.75 (–2.69)
Production shifters					
Tenure reform	0.551	0.99	54.55 (59.34)	0.00	0.00 (0.00)
Market reform	–0.0001	25.59	–0.20 (–0.28)	9.29	0.09 (0.32)
Trend	–0.035	7.00	–0.25 (–0.27)	5.00	–17.50 (–62.81)
Agroforestry					
Shelterbelts	0.001	10.84	1.08 (1.18)	4.54	0.45 (1.63)
Intercropping	0.006	6.49	3.89 (4.24)	8.23	4.94 (17.72)
Residual			2.77 (2.99)		1.16 (4.15)
Total growth			91.92 (100.00)		27.86 (100.00)

[a]Values in parentheses are percentage contributions.

younger. Agroforestry's proportional share increased to 19.4% in the second period, as the trees and their impacts matured. This second-period result supports the more casual observations of Zhong et al. (1991) and Zhu et al. (1991) of 10–15% increases in crop yields due to agroforestry. Indeed, these estimates may have been conservative for the northern plains. Most of the contribution of agroforestry in Shandong was due to intercropping, which, by itself, accounted for 4.24% of the growth in agricultural productivity between 1977 and 1984 and 17.72% between 1985 and 1990. The larger effects of shelterbelts should become apparent later, because most shelterbelts only matured after 1990.

Summary Remarks

The regression results are convincing of the proposition that conversion to the HRS was the major contributor to expanding agricultural productivity in the 1980s. It contributed to tree planting and the expansion of forestry as well, because long-term responsibility contracts for land use gave farmers the confidence that they would obtain the various improvements in agricultural productivity emanating from their own long-term conservation investments, and these investments were one source of the increase in forest cover.

The beneficial effects of trees and forests probably expanded even further into the 1990s, for two reasons:

- Trees planted in the early 1980s were approaching maturity in the 1990s, and their contribution as windbreaks was reaching its maximum. Other trees planted later were still growing, and their contributions were increasing. Farmer experience in tree and forest management also was increasing. Undoubtedly, this led to better spacing of young trees and to even more improvements in agricultural productivity.
- HRS contracts were renewed in the late 1990s and, with renewal, much of the remaining uncertainty about the continuation of the HRS policy dissipated. As a result, the climate for long-term investments in agricultural productivity improved once more. The climate for investment in the longer-term commercial products of trees and forests, products such as industrial wood, improved at the same time and for the same reason.

Conclusions

The econometric evidence convincingly supports both of the propositions from Chapter 2: improved tenure is important, but so is a stable policy environment. Tenure gives households the rights to the fruits of their investments; a secure policy environment allows households to plan and invest with the confidence that those rights will still be theirs when their investments mature.

Tenure and the relative stability of the policy environment were central features of China's forestry reforms, and they are the key predictors of the early impacts of these reforms. Undoubtedly, these two features continued to affect forest inventories well beyond the period of this analysis. They must have affected forest inventories well beyond 1990, because the long growing period for forestry means that harvest and reforestation decisions made in the early 1980s would be reflected in the numbers and volumes of 20-year-old trees in the 2000s, 30-year-old trees in the 2010s, and so on.

The evidence also supports the proposition that some of the increase in China's forest cover in the late 1980s and 1990s was designed to obtain nonmarket environmental benefits. Households planted trees to improve their agricultural environment, and they rapidly obtained benefits—even from young trees. The beneficial effects of tree planting on the ambient agricultural environment explained approximately 20% of the 5,200 kg/ha increase in agricultural productivity in the northern plains between 1978 and 1992.

Notes

1. The government's rural reforms affected farm forestry in the eastern CNNFFR and the SCFR but not the natural forests in the more arid western CNNFFR or in China's industrial Northeast or in the Southwest.

2. In a few places—notably, northwestern Fujian—an organizational form known as a *shareholding corporation* emerged. Shareholding corporations converted all collective commercial forests into monetary shares and distributed the shares to individual households. To date, shareholding corporations perform much like the old regime—only with a name change (Sun 1992).

3. The measure includes all collective forestlands. It excludes state forests because these were not eligible for household contracts. State forests are relatively more important in the South, where they account for about 10% of the total forest area.

4. Alternative specifications, (a) restricting the data to the three forest survey years and (b) dropping the trend variable but using all the data, yield similar results. They indicate that our results are robust, and they lend confidence to our conclusions.

5. The coefficients of the policy instruments are significantly different between the two regions.

6. Furthermore, the estimate of the effect of secure tenure for the northern production region is conservative because HRS and contract responsibility system reforms were largely complete in this region by 1984, but our data extend to 1989. This makes the trend coefficient more prominent relative to the coefficient on the ratio of contracted land *CS*.

7. There were no significant subsidies or regulations affecting farmer behavior. *HRS* and *CM* incorporate the major policy and institutional changes that affect farmer decisions.

8. The standard rule of thumb is that well-designed shelterbelts have a wind-moderating effect for a distance equal to twice the tree height. More mature trees are taller and their cover is denser. Density improves wind control; height extends it (Nair 1993).

9. A second examination with alternative labor data confirmed this conclusion. The State Price Administration Bureau (SPAB) collects provincewide farm labor data for specific crops. It also collects data on material costs (e.g., fertilizer, seeds, and irrigation). SPAB's labor data show a decline in farm labor for Shandong that is more consistent with the national experience. The SPAB data do not include measures of farm machinery or draft animal costs. Furthermore, their provincewide reporting scale precludes the assessment of cross-sectional variation. Applications of the SPAB labor and material data caused the labor coefficients in the regressions to become positive but it remained insignificant. The land coefficient declined, the sum of coefficients on productive inputs remained small, and the key agroforestry coefficients remained essentially unchanged from those reported in Table 4-9. These findings indicate that the general empirical observations in Table 4-9 are robust.

10. Therefore, the trend variable may capture some of the usual positive effect of technical change as well as some of the negative effect of declining infrastructure, with the latter dominating in our variable. Antle (1983) showed that a measure of infrastructure can be an important explanatory factor in agricultural productivity.

11. An unreported regression shows that the combined effects of intercropping and shelterbelts are in the neighborhood of 0.08 and significant.

References

Antle, J. 1983. Infrastructure and Aggregate Agricultural Productivity: International Averages. *Economic Development and Cultural Change* 31(3): 609–619.

China Agricultural Yearbook. 1985 through 1989. [In Chinese]. Beijing: China Agricultural Press.

China Forestry Yearbook. 1986. [In Chinese]. Beijing: China Forestry Press.

Fan, S.G. 1991. Effects of Institutional Reform and Technological Change on Production Growth in Chinese Agriculture. *American Journal of Agricultural Economics* 73(2): 266–275.

Feder, G., L.J. Lau, Y.J. Lin, and X.P. Luo. 1992. The Determinants of Farm Investment and Residential Construction in Post-Reform China. *Economic Development and Cultural Change* 40(3): 287-312.

Hayami, Yujiro, and Vernon W. Ruttan. 1985. *Agricultural Development: An International Perspective.* Baltimore, MD: Johns Hopkins University Press.

Henan Yearbook. 1986. [In Chinese]. Zhengzhou, China: People's Press of Henan.

Lin, Y.J. 1992. Rural Reforms and Agricultural Growth in China. *American Economic Review* 81: 34–51.

Lu, X.Y. 1986. *Research on Responsibility Systems* [in Chinese]. Shanghai: People's Press of Shanghai.

McMillan, J., J. Whalley, and L.J. Zhu. 1989. The Impact of China's Economic Reforms on Agricultural Productivity Growth. *Journal of Political Economy* 97: 781–807.

MOF (Ministry of Forestry). 1988. *Resource Bulletin* [in Chinese]. Beijing: Ministry of Forestry.

Nair, P.K.R. 1993. *An Introduction to Agroforestry.* Dordrecht, the Netherlands: Kluwer Academic.

Shandong Yearbook. 1985 through 1992. [In Chinese]. Jinan, China: People's Press of Shandong.

Stone, B. 1988. Development in Agricultural Technology. *The China Quarterly* 116: 762–822.

Sun, Changjin. 1992. Community Forestry in Southern China. *Journal of Forestry* 90(6): 35–39.

Wang, Y.C., D.S. Liu, and J.T. Xu. 1991. A Study of Forestry Development in the South [in Chinese]. In *Studies on China's Forestry Development,* edited by W.T. Yong. Beijing: China Forestry Press.

Wen, J.G. 1993. Total Factor Productivity Growth in China's Farming Sector: 1952–1989. *Economic Development and Cultural Change* 41: 1–41.

Yin, Runsheng. 1994. China's Rural Forestry since 1949. *Journal of World Forest Resource Management* 7: 73–100.

Yin, Runsheng, and William F. Hyde. 2000. The Impact of Agroforestry on Agricultural Productivity: The Case of Northern China. *Agroforestry Systems* 50: 179–194.

Yin, Runsheng, and David H. Newman. 1997. The Impact of Rural Reform on China's Forestry Development. *Environment and Development Economics* 2(3): 289–303.

Yin, Runsheng, and J.T. Xu. 1987. A Survey of Timber Revenue Distribution before and after Opening the Market. *Forestry Problems* 1: 109–128.

Yong, W.T. 1987. How to Enhance the Forest Productivity. *Forestry Problems* 1: 19–40.

———. 1989. Agroforestry Development and Its Implications. *Forestry Problems* 7: 127–145.

Zhong, M.G., C. Xian, and Y.M. Li. 1991. A Study of Forestry Development in the Major Agricultural Areas [in Chinese]. In *Studies on China's Forestry Development,* edited by W.T. Yong. Beijing: China Forestry Press.

Zhu, Z.H., M.T. Cai, S.J. Wang, and Y.Y. Yiang (eds.). 1991. *Agroforestry Systems in China.* Beijing and Ottawa: Chinese Academy of Forestry and International Development Research Center.

CHAPTER 5

Policy Reform and Investment in Forestry

Daowei Zhang

INVESTMENT IS AN INDICATION of the dynamism and the future potential of a sector or an economy. As China has progressed through its period of reform, a wide range of investment opportunities has arisen. This chapter reviews the trends in investment in China's forestry sector and examines the motivations for and limitations on that investment, including investment in both silviculture and the wood-processing industry and from both domestic and international sources.

In Chapter 2, D. Liu and Edmunds described the incentives that the contract responsibility system (CRS) provided for household management. CRS was initiated in the early 1980s, and China's forests began to recover in volume and area in the late 1980s. In Chapter 3, Liu et al. described the remaining disincentives for investment, with emphasis on the high tax burden on forest products. In Chapter 4, Yin demonstrated that new agricultural opportunities provided additional incentives for forest investment, but variations in the administration of the reform policies added to previous uncertainty and restrained further forest investment in some regions.

These chapters focused on the private investment opportunities that arose—mostly for farm households—over the course of approximately 20 years of China's reforms. They contained a fairly complete story for those households. However, the collectives from which those households obtained their land-use rights managed only 55% of China's forest cover and only 30% of its forest volume in 1988. These same lands still only accounted for 58% of China's forest cover and 29% of its standing volume in the most recent forest survey, in the late 1990s.

The investment experiences of the state-owned enterprises that manage almost one-half of China's forestlands and more than one-half of its standing forest volume and the motivations of the smaller numbers of private nonfarm investors and foreign investors are also important for a complete understanding of forest investment over the past 20 years. The combined record of the four sources of investment has been impressive, particularly during the 1980s, the first 10 years after the initial reforms. Ten years was a sufficient period for some new seedlings to grow to a measurable size for China's periodic forest surveys. Forest cover increased by 13.7 million ha (9.3%) and standing forest volume increased by 600 million m^3 between China's fourth and fifth forest surveys (between 1989–1993 and 1994–1998). However, 2.8 million ha of forestland were converted to other uses during this period, and 57 million ha (6% of China's total land area) remained as wasteland—suitable for forest use if an investment in afforestation were made.

This chapter provides an historical overview of these experiences. It examines each of the four categories of investors. It explores their motivations and the bases for their successes, but also inquires why so much potential forestland remains unmanaged wasteland.

My primary observations are that:

- Policy reforms in the collective sector provided some incentives for silvicultural management, but incomplete property rights, high transactions costs for these property rights, and high taxes dampen farmers' enthusiasm and restrict further silvicultural investment.
- The government's regular silvicultural investment has been modest, but project-based forest investment is increasing, largely in response to specific ecological disasters. However, some government funds have been diverted from silviculture to administration, and the costs of monitoring silvicultural activity are high.
- Private nonfarm investment in forestry is low, primarily due to the high risk associated with forestry investment in China.
- Foreign investment in China's forestry sector is small, but significant and increasing as China's economy develops.

The important questions for the future have to do with whether the recent history of gradual reforms will continue—and whether property rights will become more complete, transactions costs and taxes will become less burdensome, and the government will continue its investments in environmental improvement. If the reforms continue in the pattern of the recent past, then private domestic investments in silvicultural management will expand, and the environmental protection services of forests will expand as well. On the other hand, if the reforms in forest policy have run their course, then China's great expanse of wasteland will remain unforested, more forestland will be converted to agricultural use, and China's forests will fail to achieve their potential as environmental assets. The challenge is to continue the pattern of private economic reform and public environmental investment.

Historical Overview

Prior to 1978, China's economy was planned and highly centralized. The planned economy had two units of economic production: the state and the collectives. Investment in forestry, like investment in other sectors, came from these two sources. For purposes of this chapter, state investment includes the public expenditures of the central, provincial, prefecture, city, and county governments. The investment of the collectives includes the expenditures of township, village, and smaller production units, under the guidance of state planning policy. Since the beginning of reforms, the general public has contributed only small increments, first as voluntary citizen action to green-up the countryside and more recently as private investment in industrial forest enterprises. Foreign investment in forestry, from private sources or foreign government loans, was almost nonexistent until very recently, because China emphasized self-reliance in the pre-reform era.

The collectives typically invested small amounts in local afforestation and reforestation activities or in small wood-processing enterprises. The records of these investments are mostly in terms of numbers of trees planted or area afforested or reforested. Records of the costs of these investments are unavailable. Therefore, it is difficult to determine their effectiveness, although it is clear to most observers that the absence of any link between individual effort and individual reward generated many inefficiencies. In general, the collective sector was perceived as supplemental to the state sector, and its investments were much smaller than the state investments in forestry.

The composition of state investment has changed over the past 50 years. The central government's share of state investment in all state-owned enterprises declined gradually from 84% in 1953 to 62% in 1978 to only 5% in 1995. Meanwhile, provincial and local governments, as well as international donors, have played an increasing role. The central government's share of state investment in the forestry sector alone declined at a slower rate, from 100% in 1953 to 35% in 1995. Therefore, while the central government was reducing its role in general economic activity, it was not reducing its role in forestry nearly as rapidly. Forestry's share of total state investment (from all levels of government) increased from 1.5% in 1953 to 5.4% in 1963. Since 1963, forestry's share has declined again to 0.5% of total state investment in all sectors in 1995 (C. Zhang 1999).

State investments in forestry have included expenditures in both silviculture (forest management) and the forest industry (logging and wood processing). In the early 1950s, state investment in the forest industry dominated state investment in silviculture by a ratio of 10 to 1. After 1960, the share of state investment in silviculture gradually increased until it finally surpassed the share of state investment in the forest industry in 1988 (C. Zhang 1999). In recent years, silvicultural investments have accounted for approximately 70% of total state investment in forestry.

Farm Household Investments in Forestry

Two fundamental sets of rural reforms affected the forest investments of farm households: the devolution of property rights and the liberalization of timber markets. The transfer and improved definition of property rights improved the link between individual effort and reward, the source of low efficiency in forest use, low productivity, and low farm incomes. Liberalizing the markets increased farmers' revenues and farm income and attracted household investments in afforestation and forest cultivation, thereby increasing China's forest resources.

The markets were liberalized in two steps. First, farm households were allowed to sell production in excess of quotas on the local market beginning in 1979. Then, in 1985, the two-price system was dropped, and households were allowed to sell all their timber at market prices, although within harvest and shipment limitations established by the Ministry of Forestry (now the State Forest Administration).

Devolution occurred within the guidelines of the "Three Fix" policy, under which local governments introduced various reforms that can be generalized as contracts in the early stage of reform (1980s) and leaseholds and shareholding in the later stages (1990s). Contracts usually have a short duration (3–15 years) and are similar to profit or *profit-à-prendre*, a form of standard property right under common law. Holders of contracts have rights to profits from the management or sale of timber but not to the land on which the timber is grown. Leaseholds are for longer terms (10–50 years). Leaseholders have rights to manage, sell, and profit from trees, and the longer period of the leasehold gives them an implicit interest in the land itself.[1]

Farm households typically hold small (less than 0.2 ha) private woodlots for subsistence use. They hold contracts or leasehold rights on responsibility plots for larger areas up to hundreds or even thousands of hectares for commercial timber production. The original contracts and leaseholds were generally allocated to farmers according to the number of persons in a household or group of households. In some cases, the contracts were auctioned. Reallocation was rapid. By 1985, nearly 70% of the forestlands previously under collective control were managed by households (*China Forestry Yearbook* 1986). Subsequently, some communities (especially in Fujian) set up shareholding companies to manage their forests in preference to individual household management. Some others first distributed forests to households but later rescinded the household contracts and reassembled their forestlands under shareholding companies that are controlled by village government officials or officials elected by the shareholders.

Expectations

The impacts of market liberalization and devolution of property rights on timber prices, farm incomes, and forest investment have not been widely examined. However, these reforms are consistent with the economic conditions necessary for improvement in all three areas—prices, incomes, and investment.

The central government had held timber prices artificially low with the objective of facilitating economic development. This policy had sent an undesirable signal, and households and collectives responded by rejecting opportunities to invest in forestry. The new policy of market liberalization allowed the laws of supply and demand to operate. They encouraged investment and improved household income opportunity when and where demand was greater and supplies were scarcer.

The expectations of decollectivization are complex because of the multiple characteristics of contracts and property rights (Pearse 1990; Scott 1990; Posner 1992). Contracts such as those between the collectives and households bestow one or more of the following:

- rights to the yield or income of the property,
- rights to the use or management of the property, and
- rights to the disposal, alienation, or sale of the property.

All property rights have several characteristics, including:

- exclusiveness—the extent to which a property holder can prevent others from freely sharing in the benefits of the asset;
- duration—the time period over which the right can be exercised;
- quality of title or security—the extent to which ownership of a property right is secure, socially acknowledged, and enforceable;
- flexibility—the extent to which the specifications of a particular property right can be modified during its duration;
- transferability—the right of a property holder to transfer the entitlement to another party through sale, donation, bequest, or a similar mechanism; and
- divisibility—the extent to which the holder's management of, or income from, or risk in the property can be subdivided into smaller units or according to different uses (e.g., land and timber, multiple uses).

The first three characteristics yield benefits to holders of the rights, regardless of whether the holders are in a position to adjust the inputs they employ with the property. These characteristics facilitate better use of inputs by preventing any overlap of possession with successors or neighbors. The latter three characteristics yield benefits by allowing holders of the rights to alter their factor–land ratios. Improved freedom to modify, divide, and transfer land makes it more likely that the land will be assigned to its most productive or "highest and best" use.

The related concept of transactions costs is also important (Allen 1991). *Transactions costs* are the costs of creating and maintaining property rights, including the costs of information, negotiation, settlement, enforcement, and monitoring. Income and general welfare improve as these costs decline. Transactions costs become crucial when the political and social institutions cannot effectively control disputes about the land, illegal logging, or timber theft, as well as when government taxation and regulation remove a large share of the profit incentive for investment (e.g., D. Zhang 1999; Zhang and Flick 2001).

The critical issue for this investigation is that the new household contracts and leases bestowed some of the three rights, with some degree of each of the six characteristics. As China's experience with reform continued, contracts and leases have grown to include more of the rights and a greater degree of each characteristic. However, household property rights remain incomplete at this time, and the transactions costs associated with timber and forestland have become increasingly burdensome for many households. The property rights have improved somewhat, and some household incomes have increased as a result—but these effects have

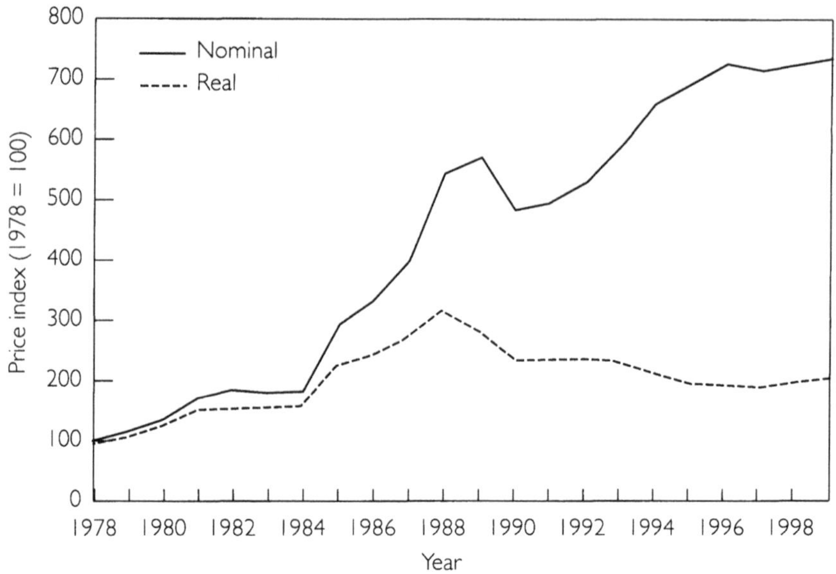

Figure 5-1. *Bamboo and Timber Price Index*

Source: China Statistical Yearbook 2000.

not been as great as they could be with even more complete property rights and lower transactions costs.

The Impacts of Reform on Farm Income and Investment

Have China's forest policy reforms improved farm incomes? Did farm households invest more in forestry as a result of these reforms?

Farm incomes have increased steadily since the reforms began in 1978, albeit at a slower pace in recent years. Specifically, per capita farm incomes increased at a nominal annual rate of 15.2% between 1978 and 1999 but at a rate of only 2.1% between 1997 and 1999 (Xiu 2001, Yang 1996). Some of the increase could be attributed to forest policy reforms, and especially to the initial transfers of collective forestlands to households and the liberalization of timber markets in the mid-1980s. Market liberalization was completed in the Southern Collective Forest Region in 1985, and real timber prices increased 43% that year (Figure 5-1).

However, soon after 1985, taxes and charges on timber began to increase more rapidly than timber prices. Since 1988, timber taxes and charges have accounted for as much as 60–70% of the sale price of logs, and the net return available to the farmer has declined to as little as 10% of the sale revenues from logs. Various reviews of farm incomes from forestry all report some level of illegal logging, a behavior that is a direct response to the perception that taxes are too high (Table 5-1). In some parts of the country, the real net farm income from timber sales was lower in the late-1990s than in the mid-1980s (FEDRC 2000).

Table 5-1. *Forest-Related Taxes and Charges as Percentage of Log Sale Price*

Source	*County or region*	*Province*	*Tax rate (%)*	*Left for farmer (%)*	*Illegal logging?*	*Impact on silvicultural investment*
Wang 1996	Kahua	Anhui	50		Yes, +15%	Negative
Chen 2000		Guangdong	34.5			Negative
		Guangxi	39.5			Negative
Z. Zhang 2000	Shanming	Fujian	76	6.8		Negative
Wang et al. 2000	Haihua	Hunan	45–55		Yes, +21%	Negative
FEDRC 1999	Jingzhou	Hunan	62	38[a]	Yes	Negative
X. Zhang 1999	Zhongyi	Jiangxi		31[a]	Yes	Negative
	Jianglue	Fujian		25[a]		Negative
Ho 1993	Jinpin	Guizhou		7.5	Yes	Negative
	Shanming	Fujian		15		Negative

[a]Including logging cost.

The answer to the second question is somewhat clearer: farmers did invest in forestry. Table 5-2 lists the changes in forest cover and standing forest volume on the collective forestlands over the course of four national forest inventories between 1981 and 1998.[2] Forest cover increased on the farm forestlands in each region and for every periodic forest survey for which the data are complete. Standing forest volume also increased on the collective lands in nearly every region, most notably beginning with the 1988–1993 forest survey. The total collective forest area for all of China increased by 19 million ha (40%) over the course of 20 years of reform. Data for standing forest volume on the collective forestlands are unavailable for the 1981 inventory, but the general opinion is that standing volume declined before it increased. Once it began increasing, the effect was sharp and rapid. Standing volume increased by 742 million m^3 (33%) between 1988 and 1998.

The earlier expansion in forest area followed by the later increase in standing volume can be explained by two factors. First, farm households tended to harvest the available mature timber as they received contract responsibility rights and leases in the early 1980s. They also reforested and even afforested some additional unforested lands. The area under forest cover increased, but the standing volume initially decreased as large volume mature timber was replaced by new seedlings and young growth. Second, forest surveys do not typically measure the volume of young trees before they attain a minimum size. Therefore, the increments in new volume were measured only in the later forest inventories once the young seedlings grew enough to become juvenile stands.[3]

These positive results notwithstanding, the quality of the collective forest is questionable. Many hectares are understocked, and large areas of wasteland have not yet been afforested. Even in the Southern Collective Forest Region, where most of the forestland has been allocated to farm households for more than 15 years, the insecurity of property rights that is embedded in the unpredictable

Table 5-2. *Timber Resources on Collective Forestlands, 1981–1998*

	1981	*1988*	*1993*	*1998*	*1988–1993*	*1993–1998*	*1988–1998*[a]
	Forest area (million ha)[a]				*Change (%)*		
Northeast	2.3	4.0	3.6	4.2	–8.2	15.4	5.9
Northwest		2.8	2.7	3.2	–3.0	17.6	14.1
Southeast	10.8[b]	12.0	13.2	16.5	9.8	25.5	37.8
Southwest		12.2	13.0	17.4	6.4	33.6	42.1
South Central		14.9	17.2	22.4	15.4	30.4	50.4
North	1.6[c]	2.1	2.5	3.4	19.3	35.0	61.0
Total		48.1	52.3	67.2	8.8	28.5	39.8
	Standing forest inventory (million m³)				*Change (%)*		
Northeast	91.2	167.3	171.1	206.7	2.3	20.8	23.5
Northwest		98.5	117.2	148.6	19.0	26.7	50.8
Southeast	595.4[b]	499.1	541.2	630.6	8.5	16.5	26.4
Southwest		883.1	917.5	1,162.1	3.9	26.7	31.6
South Central		521.9	569.5	732.4	9.1	28.6	40.3
North	29.0[c]	49.5	63.9	81.1	29.1	26.8	63.7
Total		2,219.4	2,380.6	2,961.5	7.3	24.4	33.4

[a]The standard for forested land was changed from 30% forest cover to 20% in the 1994–1998 survey.
[b]Excluding Shanghai.
[c]Excluding Tianjin.
Sources: MOF 1983, 1989, 1994 and SFA 2000.

actions of the political authorities has restrained household enthusiasm for forest investment.

The roots of this insecurity lie in forest policy and its administration. First, local governments can and do impose taxes and charges on timber without the participation and consent of the property holders. The parties involved in the contracts are not equal, and governments have imposed heavy tax burdens. For example, taxes and charges in Jingzhou County, Hunan Province, increased 170% between 1985 and 1999, while local timber prices increased only 9% and farm incomes actually decreased 84%—all in real terms (FEDRC 1999).

Approximately one-half of all forest-related taxes and charges are under the control of forest authorities in provincial, prefecture, and county governments. Many of these forest authorities have raised taxes in the name of good forest management and policy reform. For example, in the Southern Collective Forest Region (the largest farm–forest region of the country), the old price controls have been largely replaced with high taxes, many of which are under the control of forestry departments and outside the provincial budget (Zhang 1990). For example, the forest cultivation fund in Fujian Province increased 670% in real terms between 1984 and 1989. This fund is intended to support forest management. Whether it was more effective in improving forest management or in providing financial support for the agency staff is a topic for the next section of this chapter on government investment.

Second, the legal protection of property rights is weak. Infringements on property rights are a civil dispute, not a criminal offense, and boundary disputes have led to illegal logging in some regions (SFA 2001). The punishment for illegal logging and timber theft is often minimal, and it does not deter the perpetrators. Furthermore, some local governments implicitly encourage timber theft by threatening to nullify previous contracts. Yang (1996) provides an example from personal experience. He planted six hectares (private lots and responsibility plots) and invested 120,000 yuan in reforestation and management of a 150-ha collective farm under contract to him. Local government officials subsequently threatened to cancel his contract, and his forests were looted with impunity several times.

Worse yet, some local governments may have never intended to honor some contracts. For example, local officials from the Liulong Region of Shanxi Province have advocated unilateral changes in contracts following increases in the consumer price index or upon evidence that the farmer successfully generated a higher income (Liu and Bai 1996). This would increase the transactions costs for defending the property rights, and it can only discourage leaseholders from increasing their investments.

Third, China's reforms have followed a path of "crossing the river by touching stone." That is, the reforms have been gradual, and their gradual nature has created uncertainty in the minds of the land managers. The piecemeal approach to reform, combined with the prior experience from 1949 to 1979 when property rights were changed several times, has left farmers unconvinced of the direction of future policy. As a result, some farmers responded to their new assignments of property rights in the early 1980s by cutting more timber than they would have if their property rights had been more secure. (See Chapter 4 and also He 1989, Lin 1989.)

Summary

In sum, farm households have substantially extended the area and increased the volume of the forests they manage. As time has passed since the beginning of first agricultural and then forestry reforms in the late 1970s and early 1980s, some characteristics of property rights have improved. Divisibility and transferability have been assured under recent lease agreements, although some contracts and leases limit transferability to citizens of the same village or town. Durability also improved in the late 1990s, when many leases were renewed.

However, the transactions costs of forest operation remain unreasonably high and forest property rights still contain critical elements of insecurity because of

- high taxes and regulations on timber harvest levels and shipments,
- continuing boundary disputes and limited institutional protection for household contracts and leases, and

- remaining general uncertainty about the long-term stability of the general policy environment.

As a result, farm incomes may have increased, but only marginally, and households have limited their investments in forestry. At the intensive margin, households have shown a preference for agricultural and horticultural activities over forestry, and 2.8 million ha of forests were converted to other uses between the 1989–1993 and the 1994–1998 forest surveys.

Arguably, farm households do have the capital to invest in forestry. China has one of the highest savings rates in the world, and its farmers have a stock of untapped financial capital that, to this date, they have restricted to investments in improved personal housing and in the rapidly expanding local township and village enterprises. They tend to invest their labor, but not their capital, in forestry. Therefore, forest investment at the intensive margin with agriculture is less than it could be. At the extensive margin, farm households have converted some wasteland to forests, but more could be afforested if the transactions costs were lower and the property rights were more complete. Farm households do reforest after logging, although they tend to reforest with minimal effort, their silvicultural activities are not intensive, and the reforested lands remain in low-quality forests (He 1989, FEDRC 2000).

Government Investment in Forest Management

Government investment in forest management falls in two categories: regular silvicultural investments and large projects for forest protection and restoration. Funding for both passes through the state-owned forest bureaus and forest farms (state-owned forest enterprises, or SOFEs) that manage the state forestlands. Some government financing is passed on to farm households as an afforestation or reforestation subsidy.

Government Investment in the SOFEs

The two sources for investment in the SOFEs are the government and the SOFEs' own retained earnings. Both sources can be used for silviculture (forest management) or for the forest industry (logging and wood processing), and many of the SOFEs are integrated enterprises that include both in their combined land management, logging, and wood-processing operations.

Since 1978, the central government's funding has accounted for less than one-half of all investment in the SOFEs but more than one-half of the investment in SOFE silvicultural activities (Figure 5-2). (The differences are composed of provincial and local government funding and investments from the SOFEs' retained earnings.) The central government's share of all investment in the forest industry has declined with industrial reform and the "hardening" of budgets for all state-owned enterprises. Its share of silvicultural investments has increased, however, especially since 1997 and with the growing recognition of the important environmental services that forests provide.[4]

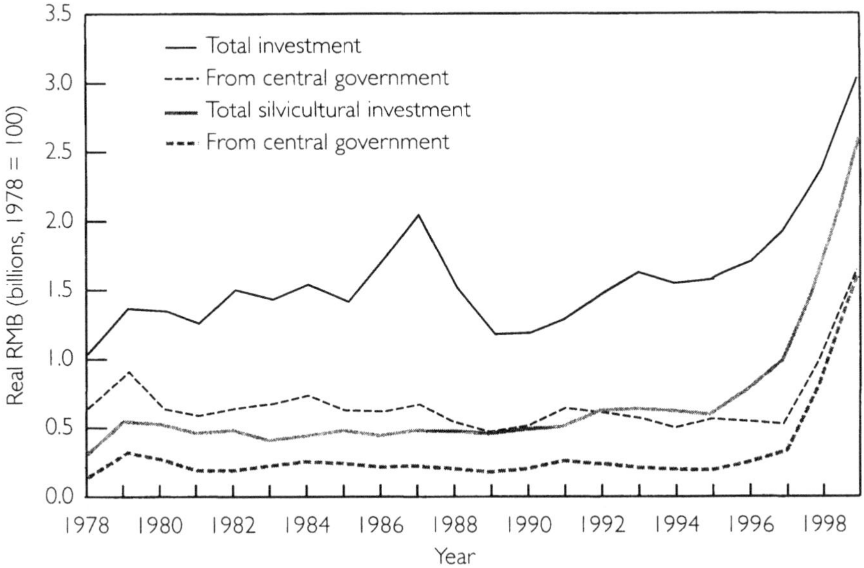

Figure 5-2. *State Investment in State-Owned Forestry Enterprises*

Source: China Statistical Yearbook 2000.

Total government investment in the SOFEs increased at a real average annual rate of 3.9% between 1979 and 1997. This aggregate rate reflects an average annual increase of 7.9% for silvicultural investments but only 2.3% for investments in the forest industry. The pattern shifted with recent decisions to leave the forest industry to the private sector and to improve the environment.

These aggregate measures mask the substantial impacts of a few large investments. The largest increases in state silvicultural investment occurred in 1979, when the Three North Forest Protection Project (Green Great Wall) was established, and between 1996 and 2001 when the government responded to several large ecological disasters (the 1998 floods of the Yangtze and Songhua Rivers and the dust storms in northern China in 2000–2001) with a decision to restrict timber harvests in some regions and to reforest and protect the upper watersheds. Another large increase in state investment in the forest industry in 1987 was probably related to the costs of controlling the 1.14 million ha fire that occurred in northeastern China that summer and the costs of reforestation after the fire. Clearly, the government of China, like those in other countries, does respond to natural crises.

Figure 5-3 traces the pattern of state investments in selective silvicultural projects since 1990. It shows that investments in large environmental protection and restoration projects have been a rapidly increasing share of the budget. There were 15 of these projects (listed in Table 5-3 along with the objective of each and its general location) before they were combined into five projects in 2001. Thirteen were ecological protection and forest restoration projects that concentrated on afforestation. One was a plantation project begun with two World Bank loans

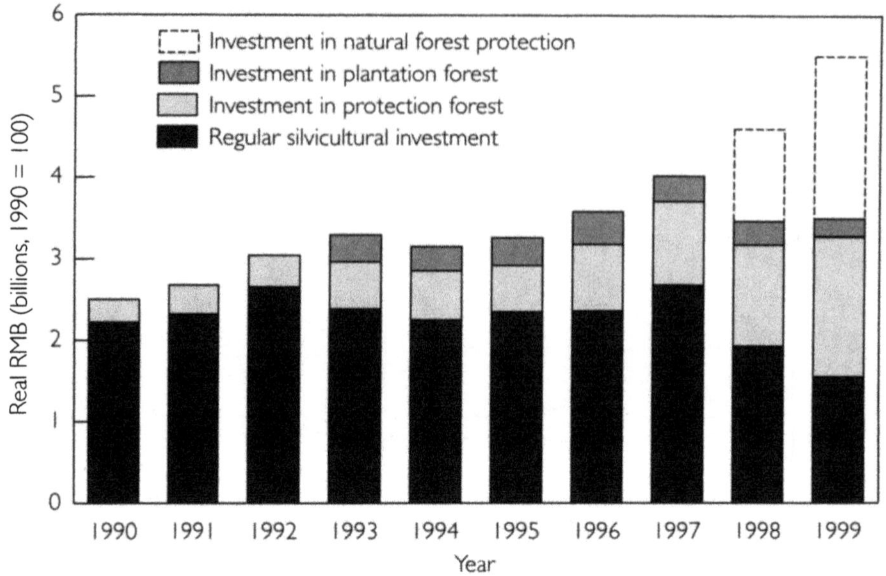

Figure 5-3. *Regular State Silvicultural Investment and Project-Based Investment*
Source: China Statistical Yearbook 2000.

to establish fast-growing forests to alleviate China's potential future timber supply problem. The last was the new Natural Forest Protection Project that began in 2000. Its focus is on the protection and restoration of existing forests. Investments in these 14 projects increased more than sixfold in 10 years.

Effectiveness

How effective were these investments? The level of state investment in forestry has probably been a smaller problem than the effectiveness and the efficiency of the investment.

The data in Figures 5-2 and 5-3 are statistical aggregates. In fact, the monies allocated for forest management do not all go to on-the-ground forestry activities. Some are dissipated in the allocation process. The forest authorities at all levels of government, as well as the SOFEs, have their own interests and responsibilities. They use their revenues to support employment, transportation, and social services for their employees, as well as for on-the-ground forest management.

Once the funding is allocated to actual silvicultural activities, its effectiveness depends on the incentive structure for SOFE employees and also for farm households, if the payment is a subsidy for farm forestry activity.

Silvicultural Investments on State Lands. As in other sectors of the Chinese economy, reforms of the SOFEs have gradually replaced a more centralized, planned system with a more decentralized, market-oriented system. Under the old system, silviculture was vertically integrated with logging and wood process-

Table 5-3. *Major State-Funded Silvicultural Projects*

Project title	*Objective and general location*	*Date of origin*	*Area (million ha)*	
			Planted	*Tended*
Three North Forest Protection Project	To build a Green Great Wall in northern, northeastern, and northwestern China that covers 45% of China's territory	1978	17.33	24.31
Central Plain Afforestation Project	To build shelterbelt forests in the region of the central plains that covers Henan, Hebei, Shendong, and Jiangsu	1984	0.42	0.62
Taihang Mountain Afforestation Project	To afforest Taihang Mountain area in northern China	1986	3.63	2.76
Coastal Protection Forest Project	To build coastal protection forests in southern provinces and protect them from hurricane invasion	1988	1.29	1.16
Upper and Middle Yangtze River Protection Forest Project	To reforest and afforest lands in upper and middle Yangtze River in southwestern and central China	1989	5.02	5.21
Plantation Forest Project based on World Bank loans	To replace fire-damaged forest with fast-growing plantation forest in northeastern China	1990	1.42	2.85
Desertification Prevention Project	To prevent the expansion of desert in northern and northwestern China	1991	1.54	0.92
Huai River Protection Forest Project	To establish protection forest in Huai River and Tai Lake watershed in central China	1996	0.20	0.32
Pearl River Protection Forest Project	To establish protection forest in Pearl River watershed in southern China	1996	0.24	0.19
Liao River Protection Forest Project	To establish protection forest in Liao River watershed in Liaoning province	1996	0.32	0.17
Yellow River Protection Forest Project	To establish protection forest in Middle Yellow River watershed in western and central China	1996	0.57	0.35
Special Ecological Protection Project	To protect area with special ecological significance in the country	1998	0.28	0.09
Returning Erosive Farmlands to Forestry Project	To convert erosive and high-slope agricultural land to forestry use	2000	0.88	0.04
Natural Forest Protection Project	To restore existing natural forests in the upper and middle Yangtze and Yellow River watersheds and northeastern China	2000	1.91	3.00

Notes: In 2001, all 14 projects were combined into five: the Beijing Area Protection Forest Project, the National Protection Forest Project, the Returning Erosive Farmlands to Forestry Project, the Natural Forest Protection Program, and the Plantation Forest Project. These five projects and the Wildlife Protection Project are now called the "big six" forest projects.

ing, and silvicultural activities were conducted by individual farms within the SOFEs or by the SOFEs themselves under a soft budget. Silviculture, logging, and wood processing were divided into separate operations in the late 1980s. In addition, the large old forest farms were divided into smaller units that now are managed by individual households or groups of households.

For silvicultural activities within the SOFEs, the first reform was the shift to a piece-rate system of wage payment. Labor is now contracted out, and wages are directly tied to the amount of work accomplished rather than the time involved (e.g., the number of seedlings removed from a nursery or the number of trees planted). This improves the market incentives for labor allocation, although the forest farms still provide general planning, materials, and machinery (Y. Zhang 2001). Arrangements of this sort are now widely used where the margins to be measured are few and the measurement costs are small.

Other payment arrangements also have developed. In one common arrangement, individuals or small teams perform specified tasks in return for a fixed payment. Most contracts stipulate minimum seedling survival rates, although there are many variations. In some cases, the forest farms provide materials and machinery. In others, the contractors possess their own material and machinery—previously purchased from the forest farms. In all cases, any profits belong to the contractors, who are responsible for organizing the labor and materials and for site preparation, planting, and tending the seedlings for the first few years of new growth.

A third arrangement, similar to sharecropping in agriculture, was developed to transfer some state-owned capital equipment to former SOFE employees. Individual workers obtain the use rights to the capital and share the income from its use with the SOFEs. This arrangement was first implemented for vehicles and small sawmills but it has spread to silvicultural and agricultural operations within the SOFEs.

A different procedure has been used to transfer use rights for land and resources. These transfers are negotiated or auctioned (especially for wastelands). The recipients of the new rights are usually limited to SOFE employees. Therefore, the land prices are not entirely determined by the market (Y. Zhang 2001). In recent years, a growing number of SOFEs have allocated lands as a substitute for salaries, pensions, and even severance pay. These payments are called "salary land," "pension land," and "employment land," respectively (CAFLU 1997).

Regardless of the specific contractual reform, the intent has been to link effort with reward, and the trend has been toward a responsibility system similar to those for agriculture and farm forestry. Y. Zhang (2001) assessed the improvements in economic efficiency due to these reforms. He concluded that the reforms reduced labor shirking and administrative costs sufficient to improve efficiency in silvicultural activities by 25% in the 40 SOFEs in Heilongjiang Province. Shen (1991) showed that, with the reforms, the three-year seedling survival rate increased from 32% in 1985 to 94% in 1990 on the Jile Forest Farm. Others (e.g., Teng and Xie 1991, Hong et al. 1992, Zhang et al. 1994) have reached similar conclusions. That is, silvicultural activities organized in a decentralized system are less costly and achieve better results than under a centralized system.

Table 5-4. *Timber Resources on State Forestlands, 1981–1998*

	1981	1988	1993	1998	1988–1993	1993–1998	1988–1998
	Forest area (million ha)[a]				Change (%)[a]		
Northeast	21.4	20.5	21.5	23.5	4.6	9.4	14.4
Northwest		5.3	4.9	5.8	–6.7	16.9	9.0
Southeast	1.3[b]	1.7	2.0	2.5	16.1	23.8	43.7
Southwest		8.2[c]	12.1	13.5	48.3	11.7	65.7
South Central		2.2	2.3	2.5	6.8	9.7	17.1
North	13.3[d]	13.1	13.5	14.2	2.8	5.3	8.2
Total		51.0	56.3	62.0	10.4	10.1	21.6
	Standing forest inventory (million m^3)				Change (%)[a]		
Northeast	2,102.8	1,980.9	2,070.0	2,151.9	4.5	4.0	8.6
Northwest		550.6	558.0	618.7	1.3	10.9	12.4
Southeast	79.1[b]	110.4	140.7	175.3	27.5	24.6	58.8
Southwest		1,594.5[c]	2,818.0	2,961.7	76.7	5.1	85.7
South Central		156.1	182.7	191.7	17.0	4.9	22.8
North	880.1[d]	901.7	936.1	1,024.9	3.8	9.5	13.7
Total		5,294.3	6,705.6	7,124.2	26.7	6.2	34.6

[a]The standard for forested land was changed from 30% to 20% forest cover in the 1994–1998 survey.
[b]Excluding Shanghai.
[c]Excluding Tibet.
[d]Excluding Tianjin.
Sources: MOF 1983, 1989, 1994 and SFA 2000.

These observations are grounds for optimism. Nevertheless, room for improvement remains. A recent China Central Television (CCTV) report (2000) shed light on the problems of monitoring silvicultural activities. The Shilinji Forest Bureau was devastated by the 1987 fire. Over the next 10 years, the bureau obtained some 27 million yuan of state funds to reforest about 9,000 ha. The reforestation was completed in 1997. However, when the project was audited, only 2,250 ha had been reforested. The CCTV report blamed official corruption, but the difficulties of monitoring within a multilayer administrative structure and the location of the bureau in the country's most isolated northern region might hide the real intentions and activities of the local officials. In a more general example, the SOFEs have reported accumulated planting over the past 20 years that is greater than the total area under their management. Yet some wastelands remain in the SOFEs. This discrepancy implies that either the reported number of planted hectares was erroneous or that many newly planted seedlings did not survive, or both. Future reforms must consider improvements in these large remaining inefficiencies.

Table 5-4 summarizes the effect of state expenditures on the forest inventory after adjustments for timber harvest removals.[5] (Table 5-4 is the state land counterpart to Table 5-2 for the collective forestlands.) It shows that the land area in state forests has remained relatively stable for all regions over the course of China's reforms. This is not surprising. Transfers of state forests to other land uses are not generally permitted. The forested area and standing forest volumes on these lands have changed, however, increasing 21% and 35%, respectively, from 1988 to 1998.

State Subsidies to Forest Farmers. State subsidies to farm households for forestry take three forms: payments for farm labor, grain support, and free seedlings. We can only speculate on the effectiveness of these transfers. When land tenure is insecure, farmers have little incentive to plant seedlings, even if the seedlings are free. When the payments for planting are in the form of grain or currency, then the extent of planting depends on the amount of the payment.

C. Zhang (1999) provides an example. She reported that a shortage of funds between 1991 and 1995 severely limited farmers' enthusiasm for the Three North Forest Protection Project. Farmers received 81 yuan per afforested hectare, an amount that was less than one-fifth of the planting cost. It was insufficient to cover even the labor opportunity cost. Zhang identified lack of funding as a prime reason that the Three North Forest Protection Project (and other projects as well) do not meet their targets for tree planting.

The available funding is often dissipated in administrative costs and, therefore, only a small share of it actually reaches the farmers. For example, in the Southern Collective Forest Region, more than one-half of all forest taxes and charges are earmarked for forestry in joint ventures involving forestry departments and farmers. However, the local and provincial forestry departments tend to institutionalize these funds and use them for their own purposes. The farmers seldom receive any monies from the government.

When subsidies are available, they may create an incentive to overreport the areas afforested or reforested. For example, Qi (1996) reported that Fuchen Village in Taihuo County, Jiangxi Province, claimed it planted trees on 40 ha. However, the actual area planted was only 22 ha, and the survival rate of the new seedlings was only 60%—far below the mandatory 85%. Shen (1991) documented overreporting of survival rates in other regions as well. Monitoring is expensive, and forestry departments do not have the means to review all lands that are reported as forested with government subsidies. Furthermore, even when the reports of planting and seedling survival are accurate, the government does not provide financial support for continued management and protection. Therefore, the incentive favors replanting rather than managing the surviving seedlings. In sum, when the lands for forests are not secured with long-term tenures, the level of management and protection declines and the newly forested areas can deteriorate to poor-quality woody brush or even wasteland within a few years.

In recent years, the authorities have extended the period of land tenure and, in some areas, they have begun to provide farmers with low-interest or interest-free loans instead of direct subsidies. These reforms may increase the farmers' interest in long-term activities and improve both initial seedling survival rates and the general effectiveness of forest protection and management.

Social and Private Nonfarm Investment

There are two sources of private nonfarm investments in forestry: voluntary public tree planting (what might be called "social investments") and private financing.

Social Investments in Forestry. China began an annual voluntary tree-planting campaign in 1981. The campaign has been responsible for the planting of 35 billion seedlings (*People's Daily* 2001). It also has raised public awareness of environmental protection. Thirty-five billion trees is a lot, but the total area planted (estimated as 5.3 million ha, assuming a plant density of 3,300 seedlings per hectare and a survival rate of 50%) is small compared with the country's remaining 57 million ha of wasteland. Furthermore, the newly planted areas tend to be more visible and accessible sites along roadways and railways. These may not be the sites in greatest need of environmental repair. Finally, the government program does not provide resources for the critical tending and protection of seedlings and growing stock that follow successful planting. Therefore, although this public tree-planting campaign has been impressive in its extent, it could do much more. And in any event, it cannot be an effective substitute for investment by the forest industry or for state investment to protect environmental values.

Private Nonfarm Financial Investments. Private nonfarm investment in forestry and the forestry industry began soon after the initial economic reforms in late 1970s. Although no official data reflect the magnitude and impact of this source of investment, anecdotal evidence and personal experience suggest that it increased rapidly in the 1990s.

Retirees and those with personal wealth or power often make private silvicultural investments. They typically return to their hometowns or to places nearby, make some arrangement with the local government to obtain land-use rights, and begin planting trees and raising high-valued animals for the local markets. For example, CCTV documented the case of a vice-president of Inner Mongolia's Textile Corporation (IMTC), who was asked to find a way to increase wool production in late 1980s. He recognized desertification as a major threat for the company's sheep herd and obtained financing from IMTC to afforest the desert fringe near the city of Baotou. Afforestation successfully prevented further desertification of the company's land. This vice-president later resigned from IMTC but continued to plant trees—relying on revenues generated from the sale of domestic animals, from loans and private donations, some government subsidy, and from volunteer labor (CCTV 2000). In another example, an executive of a private chemical company expressed interest in using the earnings of his company to green-up wasteland near Beijing.

Private investment in the forest industry has its roots in the township and village enterprises (TVEs). Prior to economic reform, the TVEs were collective enterprises. Some were small wood-processing facilities and processors of nontimber products (e.g., resin, tea oil, and bamboo roots). Soon after the reforms, some TVEs were transformed into shareholding companies, and private shareholders (often local farmers) gradually increased their shares of control at the expense of local governments. Later, nonfarmers who had accumulated capital through retail sales, transportation services, and other commercial activities recognized financial opportunity in the forest industry and began organizing their own partnerships or shareholding companies. The products of their enterprises vary, and their size ranges from very small to very large, but the number of enterprises has increased rapidly. Collectively, they produce a substantial share of China's total output of forest products. They are the most rapidly growing component of the pulp and paper industry (Xu 1999), and they account for more than 70% of the production of engineered wood products (plywood, fiberboard, and particleboard) in Fujian, Guangdong, Hebei, Jangsu, Shangdong, and Zhejiang.[6]

The purchase of equity in enterprises that produce forest products is another emerging avenue for private nonfarm investment. The performance of the shares of the larger private forest enterprises on the Shenzhen and Shanghai stock exchanges is evidence of the perceptions of private investors regarding investments in forestry. These two exchanges were organized only in 1992, but some 600 companies meet the guidelines of China's Security Exchange Commission for raising capital in these markets. Three forestry firms meet the guidelines; Yongan Forestry Limited (based in Fujian), Jilin Forestry Industry Limited (based in Jilin), and Jinggu Forestry Limited (based in Yunnan). Stock in additional firms may be traded as the economic reforms become more established.

We can examine the incentives to invest in these three enterprises with a capital asset pricing model (CAPM) of the risk and return on their stocks (Sharpe 1964, Lintner 1965). [This same model has been used to assess a wide range of industrial investments in the United States and other countries with active stock markets (e.g., Sun and Zhang 2001).] The model states that the required or expected return on an investment should be equal to the rate earned on a riskless investment plus a premium for the assumption of market risk or uncertainty:

$$R_i = R_f + \beta_i(R_m - R_f) \tag{5-1}$$

where R_i is the required rate of return on investment i, R_f is the risk-free rate of return, β_i is investment i's risk premium (commonly known as its *beta*), and R_m is the market's expected rate of return (often measured by a stock market index).

Jenson (1969) showed that CAPM is consistent with the regression equation or excess return form

$$R_i - R_f = \alpha_i + \beta_i(R_m - R_f) + \mu_i \tag{5-2}$$

where the intercept α indicates the value of an asset that is due to factors other than the market itself and μ_i are random errors. R_f is generally measured by the yield on government treasury bills. Where these data are unavailable, an alternate form of Equation 5-2 has been used:

$$R_i = \alpha_i + \beta_i R_m + \mu_i \tag{5-3}$$

A positive value for α_i indicates that the asset's expected return is greater than what the market requires for its risk class (as measured by β). Thus, a positive α_i

Table 5-5. *Capital Asset Pricing (CAPM) Results for Publicly Traded Forest Products Firms in China and the United States*

	α		β			
Asset	*Coefficient*	*t-ratio*	*Coefficient*	*t-ratio*	R^2	$\bar{R}^2$
China's forestry industry portfolio	0.0005	0.55	1.45*	27.93	0.42	0.42
U.S. forest industry portfolio						
Timberland limited partnerships	0.020	1.45	0.52*	2.80	0.15	0.13
Large forest industry company	–0.003	–0.26	1.04*	7.09	0.52	0.51
Medium forest industry company	–0.007	–0.64	0.94*	6.95	0.51	0.50

*Significant at the 1% level.

Source: U.S. forest industry portfolio data from Sun and Zhang 2001.

indicates a superior risk-adjusted return. The coefficient β_i is an indicator of the asset's market risk. If the value of this second term is greater (less) than one, the asset moves more (less) than corresponding moves by the full market. Thus, an asset with a positive (negative) beta is said to be more (less) risky than the general market.

Equation 5-3 can be applied to a portfolio equally weighted with the three publicly traded forestry companies. The market rate of return is an index composed as the simple average of the Shanghai and Shenzhen stock exchange indices. Yongan Forestry Limited was listed on Shenzhen Stock Market in December 1996; the other two enterprises were listed in October 1998 and August 2000. Therefore, the portfolio is limited, especially before October 1998, and the results of this assessment must be viewed with caution.

Table 5-5 lists the results using daily data from December 6, 1996, to May 31, 2001—as well as comparative results for U.S. forest products firms traded on the New York Stock Exchange. The risk of investing in the forest industry firms in China was 1.45 times greater than the risk for the Chinese market as a whole—over this time period. In comparison, the risk level for integrated U.S. forest products firms was only slightly higher than that of U.S. market as a whole, and the risk level for timberland investments in the United States was only one-half the risk level of the U.S. market as a whole. Therefore, the limited information on private investment in China suggests that investments in the forestry sector are more risky than the average investment in the rest of China's economy. They require a higher rate of return before investors will spend their scarce financial resources. If the U.S. comparison is representative, then investment in China's forestry sector is also more risky than investment in forestry in other countries.

These observations identify risk as a deterrent to investment, but they do not identify the sources of the additional risk. Perhaps those sources are related to uncertainties about land tenure, perhaps they are related to a generally uncertain long-term policy environment, or perhaps something else. However, a risk level greater than the general market level suggests that risk itself constrains some investment in forestry and that if the risk were reduced—that is, if the sources of this additional risk or uncertainty were eliminated—then private individuals would increase their investment in forestry. Furthermore, the U.S. evidence suggests that a high level of risk or uncertainty does not have to be a normal characteristic of the forestry sector relative to other sectors. Therefore, it should be possible to adjust policy to eliminate the abnormal risk. This will be an important task for China's continuing forestry reforms.

Foreign Investment

One of the objectives of China's reforms was to attract foreign capital. In recent years, China has attracted more foreign direct investment than any other developing country. By 2000, China had attracted US$42.7 billion of foreign capital. The forest industry itself attracted about $800 million in 1996, $900 million in 1997, and $1 billion in 1998 (Lu and Landell-Mills 2001). It accounted for more than 2% of all foreign direct investment in China in those years, a share that was more than double the share of the forest industry's contribution to national gross domestic product and larger than forestry's share of the manufacturing sector. It is also greater than, for example, the share of foreign investment in the U.S. forest industry. Therefore, one could say that China's forest sector has done well in with respect to this objective.[7]

Foreign direct investment in China comes from two sources: private firms such as Sino-Forest Corp. and Asia Pulp and Paper Ltd., and international donors such as the World Bank, Asia Development Bank, and some developed-country governments. Between 1980 and 1999, China obtained some $600 million in Word Bank loans, $250 million in foreign government loans, and nearly $300 million in foreign aid. Many of the loans and much of the foreign aid were for afforestation and reforestation. The sum of foreign direct investment in forest plantations was $100 million during this period (SFA 2001).

China's market is large. Its growth has been a source of increasing timber demand even though the country has only a limited domestic timber supply—as indicated by its increasing net imports of forest products (D. Zhang 2001). As China's per capita income continues to grow, its demand for forest products and related imports will also grow. The market's size and its growth are incentives for foreign forest investment. These have been reinforced with long-term tenures offered to foreign investors in forest plantations in southern China. However, improved tenures have not yet been especially successful in attracting investments in forest management. Rather, the foreign investments in the forest sector have been concentrated in wood processing, and especially the pulp and paper and plywood industries. These industries benefit from the lower labor costs that have also been so important in attracting investments in other components of China's manufacturing sector. China's coastal wood processors have attracted most of the foreign investment. For example, more than 85% of China's plywood production comes from eight coastal cities and provinces—Fujian, Guangdong, Jiangxu, Liaoning, Shangdong, Shanghai, Tianjin, and Zhejiang—where domestic demand is high, access to foreign suppliers of logs and machinery is good, and the re-export of products is easy.

Conclusions and Discussion

This chapter reviewed the impacts of forest policy reforms on farm household investment, government investment, private nonfarm investment, and foreign investment. The reforms have produced increases in the investments of farm households, expanding both the area under forest cover and standing forest volume, and they have improved the environment for private domestic and foreign investment. Meanwhile, the government has shifted its own forest investments from the forest industry to forest management itself and especially to management intended to improve the environmental services that forests provide.

Nevertheless, policy constraints—some remaining and some new—prevent China from achieving its forestry potential. China's large market and growing economic wealth, low labor costs, and favorable tax policies for foreign investors encourage foreign investment. However, remaining uncertainties in policy and land tenure, combined with burdensome new taxes and regulations, reduce the incentives for domestic investment in long-term forest production.

Two policy improvements are central to further enhancement of the incentives for farm households to invest in forestry: one, reducing the taxes and charges on production to a level that is related to the value of the services provided and, two, improving the property rights. Property rights could be improved by allowing the transferability of forestland and young forest stocks, lengthening tenures, and auctioning wastelands to the highest bidders. In addition, landowners should receive compensation when the government takes forests and forestlands for other uses. Knowledge that compensation is likely would increase farmers' confidence that they will reap the rewards of their long-term investments.

Providing farmers and other private investors with the right set of property rights and other incentives will probably achieve better results than increased government investment. Government investment suffers from two substantial problems:

- high administrative costs in multiple layers of government administration and the misuse of forestry funding; and
- high monitoring costs to control shirking behavior.

In the long run, controlling the size of government activity is the only way to reduce administrative costs.

In the future, China must address a growing imbalance between domestic demand and supply that could reach 25% by 2010 (Zhang et al. 1997). This will necessitate an increase in imports, encourage wood substitutes, act as an incentive to improve production efficiency, and ultimately, reduce consumption as prices rise. Two of these prospective results—imports and substitution—warrant further examination.

Wood imports may become more expensive as Southeast Asian sources are drawn down and the Russian economy recovers (and Siberian imports become more expensive). Could China increase its own production to offset rising import costs? D. Zhang (2001) pointed out that net wood import costs are four to seven times China's state investment in forestry. He also calculated that the stumpage portion of net wood product imports is approximately 60% of state silvicultural investment in any year. This means that China is already spending the financial capital that, if transferred from purchasing imports to growing domestic timber, would produce its own supply. Poor planning and inadequate policy constrain the investment.[8]

Substitution is a different option. The substitution potential of bamboo is most intriguing, but little is known about its full economic potential. China is the world's leading producer of bamboo, its bamboo production is increasing, and bamboo is a substitute for wood in almost all forest products. But what are the limits of bamboo substitution for wood, and what effects will rising wood prices and household incomes have on it? These important questions merit further inquiry.

In conclusion, China has the "hardware" of forest production—the land, labor, financial capital, and entrepreneurial skills. Its shortfalls are related to policy, in particular to taxes and regulations, secure property rights, and a stable policy environment. Improvements in each of these areas would reduce the risks and transactions costs for private forest investment. If the government addresses these issues in a satisfactory manner, it will attract private investment to commercial forest management. Government forest investment could then focus on the forest-based environmental services that private investment does not provide as well.

Notes

1. Although most contracts do stipulate that contract holders should replant trees after harvesting timber, the short-term nature of contracts gives contract holders little de facto interest in growing trees or in managing the land. Leases, on the other hand, give their holders enough time to reap the benefits of their management efforts on those lands. Therefore, leaseholds provide a de facto incentive for land and forest management.

2. In fact, Table 5-2 reflects impacts of all private investment, including private nonfarm investment and foreign private investment. However, the latter two are both very small and also recent. Therefore, farm households have been responsible for the vast bulk of direct impacts on the collective forestlands over our period of inquiry.

3. Qiao et al. (1998) supported these aggregate observations with a local examination of 28 villages in Yunnan. They found that farmers in Yunnan did invest. They afforested local wasteland and significantly expanded the area in household-managed forests.

4. The retained earnings of the SOFEs themselves are another source of silvicultural investment. The local autonomy of the industry-oriented SOFEs (mostly in the Northeast) has increased. Yet these SOFEs remain underfunded relative to the demands for their timber. Eighty percent of the SOFEs have exhausted their mature timber resources. One-third were unprofitable in 1994, and this unprofitable share rose to one-half three years later as the total indebtedness of all SOFEs rose an additional 4.5-fold. With this much indebtedness, new investment is unlikely until many conditions change. However, there are prospects for change and, as some SOFEs begin to behave more like private operations, the experiences of private nonfarm forest investors (presented in the next section) will become relevant for the SOFEs as well.

5. In fact, Table 5-4 reflects impacts of all investment on state lands, including voluntary private investment and international donor investment. However, the latter two have been small in comparison with state investment, and investment by foreign donors has been recent. Therefore, state investment, including the retained earnings of the SOFEs, has been responsible for the vast bulk of direct impacts on the state forestlands through the period of China's fifth forest inventory.

6. These six provinces account for approximately 60% of China's production of engineered wood products (Tang Xiaowen, personal communication, April 2002). The absence of data on foreign-owned enterprise makes the accurate estimation of capacity and production rates difficult.

7. However, foreign investment is concentrated in the manufacturing sector. As China opens its financial and service sectors for foreign investment, the manufacturing share of foreign investment will probably decline. Because forestry is one component of the manufacturing sector, its share of foreign investment will probably decline as well.

8. Furthermore, domestic production would have favorable effects on income distributive and regional development. It would include some production in the Northeast and the West, regions that are the focus of China's Western Regional Development Program and its Natural Forest Protection Program. Imports would encourage further development of the wood-processing industries in the coastal regions.

References

Allen, Douglas W. 1991. What Are Transaction Costs? *Research in Law and Economics* 14: 1–18.

CAFLU (China Agriculture and Forestry Labor Union). 1997. To Develop Farming in Forestry Areas in an Effective Way. *Forestry Economics* (China) 5: 25–30.

CCTV (China Central Television). 2000. Growing Forests on Report Cards. May 30.

Chen, Wanji. 2000. Recommendations on China's Forestry Development [in Chinese]. *Forest Work Research* 2: 28–36.

China Forestry Yearbook. 1986. [In Chinese]. Beijing: China Forestry Press.

China Statistical Yearbook. 2000. [In Chinese]. Beijing: China Statistical Press.

FEDRC (Forest Economics Development Research Center). 1999. An Investigation on Forestry Related Taxes and Fees: The Case of Hunan (Part I). *Forestry Economics* (China) 6: 26–35.

———. 2000. A Survey of Forestry-Related Taxes and Fees. *Forest Economics* (China) 6: 26–35.

He, Qun. 1989. Inquiry in Southern Forestry Reform. *Forestry Economics* (China) 1: 1–8.

Ho, Zhizheng. 1993. An Investigation on Forest Taxes and Fees. *Forestry Economics* (China) 1: 27–44.

Hong, H., L. Li, and G. Wei. 1992. Forestry Economic Review in Heilongjiang Province [in Chinese]. *Forestry Economics* 4: 22–24.

Jenson, M. 1969. Risk, the Pricing of Capital Assets and the Evaluation of Investment in Stock Portfolios. *Journal of Business* 42(2): 167–247.

Lin, Qing. 1989. Property Right System in Southern Collective Forestry Regions and Its Reconstruction [in Chinese]. *Forestry Economics* 6: 2–9.

Lintner, J. 1965. The Valuation of Risk Assets and the Selection of Risky Investments in Stock Portfolios and Capital Budgets. *Review of Economics and Statistics* 47: 13–37.

Liu, Tuoxin, and Jianrong Bai. 1996. Proposal to Solve Contract Disputes in Auction of Four Wastelands [in Chinese]. *Forestry Work Research* 6: 4–6.

Lu, Wenming, and Natasha Landell-Mills. 2001. Instruments for Sustainable Private Sector Forestry Project: China Country Study. Unpublished draft manuscript. London: International Institute for Environment and Development.

MOF (Ministry of Forestry). 1983. *National Forest Inventory Survey 1977–1981* [in Chinese]. Beijing: Ministry of Forestry.

———. 1989. *National Forest Inventory Survey 1984–1988* [in Chinese]. Beijing: Ministry of Forestry.

———. 1994. *National Forest Inventory Survey 1989–1993* [in Chinese]. Beijing: Ministry of Forestry.

Pearse, P.H. 1990. *Introduction to Forestry Economics*. Vancouver, British Columbia, Canada: UBC Press.

People's Daily. 2001. Some 35 Billion Trees Planted under Voluntary Campaign [in Chinese]. March 12.

Posner, R.A. 1992. *Economic Analysis of Law*, 4th ed. Boston: Little, Brown and Company.

Qi, Hong 1996. Sources of Capital and Its Utilization in Afforestation and Reforestation [in Chinese]. *Forestry Work Research* 1: 6–13.

Qiao, Fanbing, Jikun Huang, and Scott Rozelle. 1998. Forestland Tenure and Forestry Development. *Agricultural Economic Problems* 7: 23–29.

Scott, Anthony. 1990. The Market for Characteristics of Property Rights. Discussion paper. Vancouver, British Columbia, Canada: University of British Columbia, Department of Economics.

SFA (State Forestry Administration). 2000. *National Forest Inventory Survey 1994–1998* [in Chinese]. Beijing: State Forestry Administration.

———. 2001. *China's Forestry Development Report* [in Chinese]. Beijing: China Forestry Press.

Sharpe, W.F. 1964. Capital Asset Prices: A Theory of Market Equilibrium under Conditions of Risk. *Journal of Finance* 19: 425–442.

Shen, Hong. 1991. Problems Existed in Reforestation Monitoring and Recommendations [in Chinese]. *Forest Work Research* 8: 31–35.

Sun, C., and D. Zhang. 2001. Assessing the Financial Performance of Forestry-Related Investment Vehicles: Capital Asset Pricing Model vs. Arbitrage Pricing Theory. *American Journal of Agricultural Economics* 83(3): 617–628.

Teng, J., and Z. Xie. 1991. Investigation Report of Household Responsibility System in Silviculture in Yichun Region [in Chinese]. *Forestry Economics* 4: 39–42.

Wang, Xan'en. 1996. My Seeing, Hearing and Thought [in Chinese]. *Forestry Work Research* 1: 5–9.

Wang, Guang, Jinlong Liu, and Nan Xang. 2000. An Investigation on Forestry Taxes and Fees in Huihua Region [in Chinese]. *Forestry Work Research* 9: 35–46.

Xiu, Lianzhong. 2001. Factors Influencing Change in Farmers' Income. *Problems of Agricultural Economics* 5: 20–29.

Xu, J. 1999. China's Paper Industry: Growth and Environmental Policy during Economic Reform. Ph.D. thesis. Blacksburg: Virginia Polytechnic Institute and State University.

Yang, Zhichai. 1996. Please Protect My Lawful Income [in Chinese]. *Forest Work Research* 3: 63–66.

Zhang, Caihong. 1999. Forestry Investment and Forestry Economic Growth [in Chinese]. Ph.D. thesis. Beijing, China: Beijing Forestry University.

Zhang, Daowei. 1999. Endangered Species and Timber Harvesting: The Case of Red-Cockaded Woodpecker. Working paper, Auburn University. Forthcoming in *Economic Inquiry*.

———. 2001. Why So Much Forestland in China Would Not Grow Trees? [in Chinese]. *Management World* 3: 120–125.

Zhang, Daowei, and Warren Flick. 2001. Sticks, Carrots, and Silvicultural Investment. *Land Economics* 77(3): 443–456.

Zhang, Lang. 1990. Tax Policy and Its Implementation in Southern Collective Forestry Region: Analysis and Consideration [in Chinese]. *Forestry Work Research* 10: 6–13.

Zhang, Xiaojing. 1999. The Forestry Taxation and Charges Issue in the Collective Forestry Area of Fujian and Jiangxi Provinces. *Forestry Economics* (China) 6: 36–45.

Zhang, Yaoqi. 2001. The Impacts of Economic Reforms on the Efficiency of Silviculture in China: A Non-Parametric Approach. *Environment and Development Economics* 7(1): 107–122.

Zhang, Yibing, Joseph Buongiorno, and Dali Zhang. 1997. China's Economic and Demographic Growth, Forest Products Consumption, and Wood Requirements: 1949 to 2010. *Forest Products Journal* 47(4): 27–35.

Zhang, Z., Y. Li, and F. Du. 1994. A Good Forest Protection Approach: Contracted Responsibility System [in Chinese]. *Heilongjiang Forestry Survey* 1: 8–10.

Zhang, Zhefu. 2000. Distribution of Forest Income Must Be Adjusted [in Chinese]. *Forestry Work Research* 8: 42–45.

CHAPTER 6

Forest Exploitation and Protection in Reform China

Assessing the Impacts of Policy and Economic Growth

Scott Rozelle, Jikun Huang, and Vince Benziger

PREVIOUS CHAPTERS REVIEWED the primary reform policies that had direct effects on forestry and the forest policy restrictions that constrain further development of the sector. This chapter and the next have a broader responsibility. We are interested in the influences of forest policy, but we are also aware that agricultural policy and even broader macroeconomic policy and general economic growth affect the forest sector. Our objective in this chapter is to assess the relative effects of all policy and market developments that have affected China's forests since 1978—direct forest policies as well as policies and market effects that have spilled over to forestry from other sectors and from the economy as a whole.

For the forests themselves, China's leaders tried to create incentives for efficient management by designing two series of policy measures targeted at encouraging afforestation and limiting excessive harvests. One set was directed at the management practices of state-owned forest enterprises, including the forest bureaus that carried out much of the logging and transport activities in China's forests during the 1950s, 1960s, and 1970s (Ross 1988). The second set was directed at the actions of the farmers and other private forest managers who gained greater access to local forests in the early 1980s (Sun 1992; Yin 1994).

Although the reforms in most other sectors have been lauded as successful (e.g., Lin 1992, McMillan and Naughton 1992, Qian and Xu 1993, Jefferson and Rawski 1994, Qian 1994), there is disagreement about the achievements of reforms for the forestry sector. *China Statistical Yearbook* (1989) identifies a nationwide increase of 8.5% in forest cover between 1980 and 1988 (Table 6-1), and

We gratefully acknowledge comments by Runsheng Yin, Jintao Xu, William Hyde, Jim Harkness, Guo Li, Heidi Albers, Scott Pearson, Rick Scoby, and Li Zhou. The Ford Foundation in Beijing and the MacArthur Foundation provided financial support.

Table 6-1. *Forest Cover and Standing Forest Volume, Late 1970s to Early 1990s*

	Forest cover					Standing forest volume				
	%			% Change		Million m^3			% Change	
Region	1980	1988	1993	1980–1988	1988–1993	1980	1988	1993	1980–1988	1988–1993
Northeast State Forest Region										
Heilongjiang	33.6	34.4	35.6	2.2	3.5	14.4	13.2	13.5	-8.3	2.3
Nei Monggol	11.9	11.9	12.1	0.3	1.7	8.5	8.6	9.0	2.0	3.7
Jilin	32.2	33.0	33.6	2.5	1.9	6.6	7.1	7.6	8.1	6.8
Subtotal	19.5	19.8	20.3	1.6	2.5	29.4	28.9	30.0	–1.7	3.8
Southern Collective Forest Region										
Fujian	37.0	41.2	50.6	11.3	22.9	3.0	2.6	3.2	–11.0	21.9
Jiangxi	32.8	35.9	40.4	9.6	12.3	2.4	1.7	1.8	–28.7	7.3
Hunan	32.5	31.9	32.8	–1.9	2.9	1.6	1.4	1.5	–12.2	7.7
Guangdong[a]	27.7	27.0	35.9	–2.4	32.8	2.0	1.9	2.2	–8.7	18.1
Guangxi	22.0	22.0	25.3	0.0	15.2	2.2	2.0	2.1	–7.5	4.7
Subtotal	29.4	30.2	35.2	2.8	16.5	11.2	9.6	10.9	–13.8	12.9
Southwest Mixed-Management Region										
Yunnan	24.0	24.4	24.6	1.6	0.8	11.0	11.0	11.0	–0.0	0.8
Sichuan	12.0	19.2	20.4	60.1	6.0	10.5	12.7	13.0	21.4	2.5
Subtotal	16.9	21.3	22.1	26.2	3.6	21.5	23.7	24.1	10.4	1.7
North Central Fast-Growth Region										
Jiangsu	3.2	3.8	4.0	17.5	6.9	0.03	0.07	0.08	113.1	18.2
Anhui	13.0	16.4	16.3	25.8	–0.2	0.55	0.71	0.63	30.9	–12.5
Shandong	5.9	10.5	10.7	77.8	2.0	0.05	0.11	0.15	118.3	42.1
Henan	8.5	9.4	10.5	10.7	11.6	0.32	0.40	0.48	26.8	19.2
Subtotal	7.9	10.4	10.8	30.9	4.1	0.95	1.29	1.34	36.8	3.5
All China	12.0	13.0	13.9	8.2	7.2	79.8	80.9	90.9	1.4	12.3
Total covered area (million ha)	115.3	124.7	133.7							

Notes: Forest cover is total forest area divided by total land area. The years 1980, 1988, and 1993 are approximate dates of China's second, third, and fourth forest surveys. (The first survey occurred before reforms began in 1978 and is irrelevant to our analysis. The second was conducted in different provinces between 1978 and 1981, the third was conducted between 1986 and 1989, and the fourth was conducted between 1992 and 1994.)

[a]Hainan is included with Guangdong.

Sources: National Forest Resource Statistics 1983, 1989, 1994.

the World Bank (1992) points to the efforts of the forest ministry as one of the more successful examples of China's early environmental effort. In contrast, others warn of the consequences of the irreversible destruction of China's natural forests, including the loss of diverse flora and fauna (Smil 1984, 1993; He 1991; Harkness forthcoming) and the rapid deforestation of ecologically important areas such as Hainan (Schur 1984). The government's recent decision to reorganize the forest ministry may be a sign of the central leadership's dissatisfaction with the sector's performance.[1]

In fact, many of these opinions rely on broad national data on forest cover, and those aggregate data obscure critical adjustments in important classes of forests and regions of the country. In our opinion, little hard evidence or broad-based research supports either the loud cries of dissatisfaction or the bold assertions of successful transformation of either the state or the collective forestry sector. We intend to address some of the obvious questions: What is the nature of the change of forest resources? As forest cover has increased, what has happened to the structure of the forest and the quality of the resource? How do these changes affect the quality of forest-based environmental services such as soil protection and the provision of unique habitat? What are the sources of the changes in forest cover and the standing forest stock? Has the state or the collective sector contributed more to or detracted more from the development of new forests and the protection of forest resources? Has national policy been effective?

We will begin our assessment by documenting the changes in China's forest resources and by decomposing them regionally, by class of forest use, and into natural stands and plantations. We will also examine a number of economic trends, and the management reforms and policy decisions that may have contributed to these trends. Our observations provide the basis for a multivariate model for evaluating the differential impacts of various reform policies and market variables on China's state and nonstate forests between 1980 and 1993. At the close of the chapter, we will discuss the remaining problems that pose challenges for policymakers, and present our own suggestions for increasing the effectiveness of forest development and forest protection.

Our analysis relies on the standard national forest inventory data collected in the late 1970s (henceforth 1980), the mid- to late 1980s (henceforth 1988), and the early 1990s (henceforth 1993). These same data underlie the discussions in Chapters 1, 9, and 10 and the analyses in Chapters 2, 4, 5, and 7 in this book. Therefore, our observations should be consistent with the observations in these other chapters, even while our analysis is different in the policy and market influences we seek to understand.

Forest Resource Census Data and Management Types

Survey Design

Our data originate with three national forest surveys conducted in the late 1970s, the mid- to late 1980s, and the early 1990s (*National Forest Resource Statistics* 1983,

1989, and 1994, respectively). Twenty-eight provinces undertook comprehensive forest surveys under the coordination of the Ministry of Forestry (MOF). The survey design for these surveys was comparable to the forest surveys of many western and developed countries. The combined surveys create a "continuous forest inventory" that provides consistent measures of forest area and volume from one survey period to the next.

The foundation of these three surveys is a comprehensive spatial sample that relies primarily on direct observation to produce statistically valid observations at the provincial level. In Yunnan, for example, the survey organizers randomly selected 7,975 plots of 0.08 ha located throughout the province. Of these, about 30% contained forests. The remaining 70% were on cultivated land, in urban areas, in lakes, or in other nonforested regions. Trained technicians staked each forested plot with permanent markers, and enumerators visited the same plot in each succeeding survey. Officials in one province estimate that more than 80% of their survey personnel participated in both 1980s surveys. Yunnan forest bureau officials claim that more than 95.5% of the plots surveyed in the late 1970s were located and resurveyed in 1988.

The forest technicians collected area and volume data from each sample plot, and tabulated these data by age structure, tree species, and land-use category. Each province accounted for region-specific environmental factors by generating its own criteria for variables such as stand age and species composition, but national MOF coordination teams approved all provincial criteria. Supervisors monitored the work of the survey technicians and checked it for accuracy and consistency. The MOF collated and published the survey results in Beijing. The survey data do not suffer the potential bias of local and regional officials who have their own output targets and other objectives to defend because these data were never subjected to the inspection and approval of the usual nonforest government hierarchy.[2]

Forest Regions and Management Types

China categorizes its forests by four geographic regions and two management regimes. Two institutions, the state and the collectives, have shared control of China's forests since the 1950s (Ross 1988, Sun 1992, Yin 1994). In the 1970s, the state managed about 40% of China's total forest area, including the vast majority of its high-quality timber reserves. Approximately 4,000 state-owned forest farms, employing more than 500,000 workers, conducted most of the production activities (*China Forestry Yearbook* 1987, Yin 1994). Most forest farms reported to local forest bureaus, which prior to 1997 were the lowest level of the MOF command hierarchy (Richardson 1990).

State forest farm management is most closely associated with the forests of China's three northeastern provinces (Heilongjiang, Inner Mongolia, and Jilin), where state farms controlled more than 91% of the forest area in 1988 (*National Forest Resource Statistics* 1984–1988). These provinces contain China's richest industrial timber resources. In 1988, they contained approximately 39% of China's industrial forest area and 42% of its standing forest volume. They supplied approximately one-half of China's industrial timber production (Table 6-1, *National Forest Resource Statistics* 1984–1988; *China Statistical Yearbook* 1991).

Collective forest management began in the mid-1950s, when 96% of rural land was placed under the jurisdiction of village (brigade) leaders (Richardson 1990). During the 1960s and 1970s, national grain policy and party ideology discouraged local households and collectives from investing in forests, and state-owned forest farms frequently encroached on the forests of the local collectives and mined their timber resources. Genuine collective management of the forests was restored in the 1980s as part of the agricultural reforms. Village leaders and other individuals in the villages now exercise the responsibility for forest management, although these managers must answer to officials from the local forest bureaus.

Southern and north central China are the major regions for collective forest operations (Table 6-1). The southern collectives managed 89% of the forest area in that region (*National Forest Resource Statistics* 1984–1988) and they harvested approximately one-fourth of China's industrial timber in the mid-1980s (*China Forestry Yearbook* 1992, *China Statistical Yearbook* 1993). The collectives' share of timber harvests nationwide was approximately 40%. It has increased, and the state sector's share of total timber production has fallen sharply, since the late 1980s.

The north central region emerged as a major new source of industrial wood in the 1990s. Forest management in this region is similar to that in the Southern Collective Forest Region, although the north central region is less well-endowed with natural forests and its leaders have only begun to encourage the collectives to engage in forestry activities on a large scale. The prevalence of fast-growing timber species and rapidly growing young stands give this region its identity as the "fast growth" region.

Two major forestry provinces in the southwest, Yunnan and Sichuan, are more difficult to categorize. Ross (1988) considers forests in these two provinces to be state run. On the other hand, these provinces have the largest collective forestry sector in China outside of the five southern provinces (*China Statistical Yearbook* 1985 through 1993), and Sun (1992) groups them with the provinces where collective management is more common. In fact, the forests of Yunnan and Sichuan have characteristics of both management types, with a nearly equal division of forested area between the two management systems. Hence, like Yin (1994), we will treat Yunnan and Sichuan as a separate category of mixed management. Industrial timber harvests from state forest farms in Sichuan and Yunnan ranked fourth and ninth, respectively, among China's provinces in 1991. The collectives in these two provinces harvested a comparable volume. The combination of the state forest farms and the collectives of the two provinces supplied a little more than 10% of China's industrial timber (*National Forest Resource Statistics* 1991).

To summarize, China contains four major regions for forest management. The primary management regimes identify the regions and the 14 provinces that fall within them:

- Northeast State Forest Region—three northeastern provinces where the forests are predominantly state owned;

- Southern Collective Forest Region—five provinces in the south where the collective sector is dominant;
- North Central Fast-Growth Region—four provinces where collective forest management is a newly important activity; and
- Southwest Mixed-Management Region—two provinces, Sichuan and Yunnan, where management has characteristics of both modes.

These 14 provinces contained more than 75% of China's total forestland and supplied 88% of its annual timber harvests in the late 1980s (*China Agricultural Yearbook* 1989). In addition to the timber resource, these four regions contain some of the China's largest shelterbelt projects (forests planted primarily to combat soil erosion caused by wind and water) and some of the largest forest reserves and forest recreational areas (World Bank 1992, Ghimire 1994).

Forest Resources in the Reform Era

Table 6-1 shows that China's forest cover and forest volume expanded substantially during the reform era, although the pace of expansion differed from the earlier reform period between the 1980 and 1988 forest surveys and later reform period between the 1988 and 1993 surveys. Forest cover, measured as the proportion of all land with at least 30% coverage by forests and commercial tree crops, expanded steadily throughout the entire reform period, growing from 12.0% of China's total land area in 1980 to 13.0% in 1988 to 13.9% in 1993, an increase of almost 15% over the 15-year period. The land area in forests and tree crops increased by about 9 million ha in each forest survey period, a total increase of 18 million ha (from 115.3 million ha in 1980 to 133.7 million ha in 1993). Although the percentage increases in forest cover varied among China's main forest regions, all four regions experienced net growth.

The expansion of standing forest volume was more variable over time and across regions. Forest stock was nearly stagnant during the earlier reform period, increasing only from 79.8 to 80.9 million m^3, an annual rate of 0.14%. In the later reform period, however, stock increased sharply from 80.9 to 90.9 million m^3, an annual rate of 2.4%. Growth varied significantly between regions. The Northeast State Forest and Southern Collective Forest Regions experienced declines in standing stock in the early period that were offset by rapid growth in the Southwest Mixed-Management and North Central Fast-Growth Regions. Standing stock increased in all regions in the later period, led primarily by the new growth in the south.

Forest Cover

What kinds of forestry activities accounted for the 18 million ha increase in aggregate forest cover? Unfortunately, accounting for the large expansion in total forest area is not a straightforward procedure. China's forest statistics include natural forests and forest plantations. They also include vast areas in shelterbelts, largely designed to control wind erosion, and commercial tree crops such as fruit, bamboo, and rubber that would not appear in the forest statistics of most countries. The published forest inventory statistics permit some distinctions among these categories of forest, but they were not designed to address questions about reforestation and new forest plantations. For example: Did new forests replace old forests on the same land area, or were new forests planted in areas that previously were either wasteland or cultivated agricultural land? Where natural forests were harvested, were they replaced by natural regeneration, plantations of the same forest type, or plantations of exotic species? What was the survival rate for the young seedlings, and what is the stocking level of the new forest? We must rely on a mixture of national forest inventory data, other published sources, and several assumptions to answer these questions, yet some of the answers will remain uncertain despite our best efforts.

Categories of Forest Cover. China's official forest statistics include the land area in natural forests, commercial timber plantations, shelterbelts, and what China identifies as "economic" tree crops (e.g., orchards, rubber, and bamboo). The area that most of the world generally thinks of as forest (i.e., natural forests plus timber plantations) makes up the largest share of China's forestlands. It expanded only slowly during the reform period. It stagnated between 1980 and 1988 (at 80 million ha), then increased approximately 5% between 1988 and 1993 (to 84 million ha).

A relatively constant land area under forest cover, however, does not mean that reform policies, management practices, and other economic forces have not affected the resource. Table 6-2 documents the rapid increase in "engineered" forest plantations (*gongchengzaolin*) that occurred throughout the reform period. China defines *engineered forests* as plantations that were planned before the planting occurred and managed afterward. They are often supported by state funding, and their general objective is the afforestation of previously barren lands, although in some cases engineered plantations occur on lands that were selectively harvested of their mature commercial timber and then replanted with preferred species. From 1980 to 1993, both the state forest farms and the collectives continuously increased the area of newly afforested lands. The combined increase for the two management types averaged approximately 1.5 million ha/year, until the increase totaled almost 21 million ha by 1993. This increase exceeds the entire net increase of 18 million ha for all forests, including shelterbelts and tree crops. An unidentified additional increment of nonengineered reforestation occurred largely on the lands of collective farmers.

This characterization of the newly afforested area can be refined somewhat. Since 1987, the official data on engineered forests have distinguished the share that is "fast growing." Fast-growing forests are defined as forests that increase in volume by at least 15 m^3/ha/year. These have expanded by almost 0.5 million ha annually. According to Yin (1994) and species data from the national forest survey, most of the expansion occurred in the collectively managed southern and north central regions. Fast-growing plantations were often grown to support the rapidly growing paper industry and the light industrial manufacturing sector. However, despite

Table 6-2. *Newly Afforested and Reforested Plantations, 1980–1993*

Year	Afforested area (million ha)		Reforested area (million ha)
	Total[a]	Fast growing	
1980	0.91	—	0.37
1981	0.93	—	0.37
1982	0.96	—	0.38
1983	1.01	—	0.42
1984	1.09	—	0.48
1985	1.15	—	0.54
1986	1.09	—	0.49
1987	1.23	0.38	0.59
1988	1.66	0.45	0.54
1989	1.63	0.42	0.57
1990	1.91	0.45	0.54
1991	2.26	0.56	0.54
1992	2.53	0.62	0.57
1993	2.28	0.51	0.61
Total	20.74	3.39	7.00

Notes: Afforested area ("engineered" afforestation [*gongchengzaolin*]) includes total planned area and the area afforested with fast-growing species (*shushenglin*). The latter has been formally recorded since 1987.
[a]Estimates for 1980 and 1986 are extrapolated from data for 1987–1993.

Sources: China Forestry Yearbook 1986 through 1993.

the high profile of fast-growing stocks within China's forestry community (and indeed, within the international forestry community), the area in plantations from native stock expanded much faster, and native stock accounts for a larger share of engineered afforestation for each year that records are available.

Shelterbelts and "Economic" Tree Crops. China's effort to combat erosion and desertification is well documented (Richardson 1990; World Bank 1992; Rozelle et al. 1998). It is one of the largest targeted afforestation efforts in history. The central government's leaders have placed great emphasis on high-profile projects such as Inner Mongolia's "Green Great Wall" and the Upper/Middle Yangtze River Valley Afforestation Campaign. Indeed, a large portion of central government's budget for the rural environment and natural resource protection funds afforestation projects, and the most common use of corvée labor (after maintaining village roads and irrigation canals) is for tree planting.[3]

The support of national leaders and the hard work of farmers have made the shelterbelts a key component of forest cover. Shelterbelts expanded by more than 4 million ha in the earlier reform period and by 1.5 million ha during the later period—a total of 6.07 million ha between 1980 and 1993 (Table 6-3). Most of the early expansion came in China's key forest regions, all of which exhibited increases from 1980 to 1988. In the later reform period, however, only the

Table 6-3. *Shelterbelts: Area and Volume for Selected Regions, Late 1970s to Early 1990s*

	Area					Volume				
	Million ha			% Change		Million m³			% Change	
Region	*1980*	*1988*	*1993*	*1980–1988*	*1988–1993*	*1980*	*1988*	*1993*	*1980–1988*	*1988–1993*
Northeast State Forest Region										
Heilongjiang	0.47	0.45	0.40	–6	–10	0.42	0.29	0.28	–32	–3
Nei Monggol	0.28	0.40	0.50	43	26	0.12	0.21	0.23	76	8
Jilin	0.89	0.82	0.75	–8	–8	0.70	0.63	0.79	–10	26
Subtotal	1.64	1.66	1.65	1	0	1.24	1.13	1.30	–9	15
Southern Collective Forest Region										
Fujian	0.14	0.22	0.31	60	39	0.09	0.12	0.22	41	78
Jiangxi	0.05	0.19	0.26	276	33	0.04	0.05	0.07	27	35
Hunan	0.04	0.16	0.20	258	22	0.02	0.09	0.10	307	17
Guangdong[a]	0.17	0.25	0.36	46	43	0.08	0.16	0.20	92	24
Guangxi	0.59	0.67	0.86	14	29	0.46	0.47	0.57	1	22
Subtotal	0.99	1.50	1.98	50	32	0.70	0.89	1.17	28	30
Southwest Mixed-Management Region										
Yunnan	0.54	1.60	1.80	198	12	0.97	2.33	2.57	140	10
Sichuan	2.03	3.38	2.72	67	–20	2.68	5.30	5.37	97	1
Subtotal	2.57	4.98	4.52	94	–9	3.65	7.63	7.94	109	4
North Central Fast-Growth Region										
Jiangsu	0.03	0.06	0.07	115	13	0.01	0.02	0.03	222	23
Anhui	0.03	0.39	0.12	1,450	–70	0.00	0.19	0.04	4,581	–80
Shandong	0.47	0.47	0.44	1	–7	0.03	0.05	0.07	52	50
Henan	0.28	0.34	0.31	23	–8	0.13	0.15	0.16	15	11
Subtotal	0.80	1.27	0.94	59	–26	0.17	41	0.30	138	–26
All China	10.00	14.56	16.07	46	10	8.84	14.00	17.78	58	27

Notes: The years 1980, 1988, and 1993 are approximate for the three forest survey periods.
[a]Hainan is included with Guangdong.
Sources: National Forest Resource Statistics 1983, 1989, 1994.

Southern Collective Forest Region increased its area in shelterbelts (by 32%). The increase in China's overall shelterbelt area between 1988 and 1993 extended beyond the traditional forest regions to other provinces suffering severe erosion. It reflects the government's intention to expand the geographic scope of environmental protection.

The expansion of economic tree crops is one of the more noticeable changes in rural China today. These are oil-bearing trees, fruit and nut orchards, rubber, and other cash-producing nontimber tree species. The land area in these crops increased more than 40% over the full reform period (Table 6-4). Orchards accounted for nearly 80% of the increase, even though the land area in orchards made up only 19% of the area in economic tree crops in 1980. Orchards expanded by 127% between 1980 and 1988 and by 55% more by 1993. The land area in orchards increased in each forest survey and for all provinces except Jilin between the 1988 and 1993 surveys.

Accounting for Changes in Forest Cover

Table 6-5 provides a stylized account of the 18 million ha increase in forest cover between 1980 and 1993. The land areas in shelterbelts and economic tree crops expanded by 6 million and 5 million ha, respectively. Approximately 2 million ha were newly regenerated for fuelwood forests in natural reserves and in areas of other special uses. The remaining 5 million of the 18 million ha expansion was in timberland and remote forests that could have been reforested for other reasons. Because the net increase due to engineered afforestation or reforestation was 14 million ha, there must have been a net loss of at least 9 million ha of natural forest that were cleared and left without successful regeneration or converted to agricultural uses.[4]

This reconciliation of the forest statistics shows that China's forest structure evolved radically during the reform era. It suggests a critical transition in the composition of the forest and a substantial modification in the provision of forest-based environmental services. China had experienced only very limited reforestation during the 25 years preceding the beginning of reforms in 1980 (Ross 1988). The forest that remained was almost entirely a natural forest. The area in forest plantations was limited, and virtually none of the plantations were mature. Therefore, the remaining forest in 1980 was largely composed of degraded natural forest or sparse and inaccessible material.

Following the initial reforms of the early 1980s, at least 9 million ha were deforested without replacement (Table 6-5). An additional 7 million ha were logged and reforested but, in the 1980s, this reforested area contained only seedlings and juvenile growth plus the residual stock of lower-valued species and size classes that had not been removed in earlier harvests (Table 6-2). Some of the 7 million ha were replanted with native stock, and some with fast-growing stock. Almost 21 million ha were reforested or afforested for the first time between 1980 and 1993. This implies a tremendous loss of natural forest, a 7 million ha shift from natural forests to engineered stands, and a large shift in the balance of age classes of the remaining and new forest toward juvenile growth, some of

Table 6-4. *Area in "Economic" Forest, Late 1970s to Early 1990s*

	Commercial tree crops					Orchards				
	Million ha			*% Change*		*Million ha*			*% Change*	
Region	*1980*	*1988*	*1993*	*1980–1988*	*1988–1993*	*1980*	*1988*	*1993*	*1980–1988*	*1988–1993*
Northeast State Forest Region										
Heilongjiang	0.03	0.06	0.05	98	–15	0.01	0.01	0.02	0	197
Nei Monggol	0.85	0.89	0.87	5	–3	0.02	0.03	0.04	17	59
Jilin	0.05	0.04	0.04	–13	–2	0.02	0.02	0.02	18	–33
Subtotal	0.94	1.00	0.96	7	–4	0.05	0.06	0.08	15	35
Southern Collective Forest Region										
Fujian	0.34	0.57	0.79	65	40	0.04	0.20	0.44	460	115
Jiangxi	0.98	1.10	1.13	12	3	0	0.04	0.05	—	40
Hunan	2.42	2.42	2.27	0	–6	0.90	0.83	1.79	–7	116
Guangdong[a]	0.75	0.84	1.32	11	57	0.20	0.19	0.50	–5	163
Guangxi	0.71	0.80	0.99	13	23	0.05	0.06	0.20	30	223
Subtotal	5.20	5.73	6.50	10	13	0.37	0.57	1.36	54	138
Southwest Mixed-Management Region										
Yunnan	0.32	0.59	0.68	84	15	0.05	0.09	0.14	62	61
Sichuan	0.24	0.69	0.84	191	21	0.02	0.25	0.31	1,188	25
Subtotal	0.56	1.28	1.52	130	18	0.07	0.34	0.45	361	34
North Central Fast-Growth Region										
Jiangsu	0.11	0.15	0.16	27	11	0.05	0.08	0.08	70	6
Anhui	0.14	0.32	0.41	125	28	0.03	0.07	0.12	133	89
Shandong	0.33	0.87	0.99	160	14	0.17	0.73	0.86	341	18
Henan	0.31	0.33	0.43	6	31	0.10	0.10	0.17	–3	77
Subtotal	0.90	1.67	1.99	84	20	0.35	98	123	185	28
All China	11.28	13.74	16.10	22	17	1.50	3.41	5.30	127	55

Notes: The years 1980, 1988, and 1993 are approximate for the three forest surveys.
[a]Hainan is included with Guangdong.
Sources: National Forest Resource Statistics 1983, 1989, 1994.

Table 6-5. *Change in Total Forest Area during the Reform Era, 1980–1993*

Forest category		*Change (million ha)*
Total change in forest area (see Table 6-1)		+18
Accounted for by		
"Classic" forest (from *National Forest Resource Statistics* 1983, 1989, 1994)		+5
Engineered afforestation and reforestation (from Table 6-2)	+14[a]	
Land that was deforested and removed from the forest base (calculated)	–9[b]	
Shelterbelts and "economic" tree crops		
Shelterbelts (from Table 6-3)		+6[a]
"Economic" tree crops (from Table 6-4)		+5[a]
Other forest categories:		
Fuelwood forests, natural reserves, other forests (from *National Forest Resource Statistics* 1983, 1989, 1994)		+2[c]
Other changes affecting forest diversity (but not forest cover)		
Removal of natural forest cover	–7[b]	
Replanting as engineered forest (Table 6-2)	+7[a]	

[a]Mostly government-funded afforestation of wasteland.
[b]Mostly diverse natural or second growth forest.
[c]Partly planted and partly natural reforestation; actual increase is 2.6 million ha.

which was so young that it would not grow to sufficient size to appear in the next forest survey.[5]

The shifts in forest category and age structure have major consequences for the diversity of China's flora and fauna. Deforestation is correlated with a loss of biodiversity, younger stands do not typically act as repositories for the same plant and wildlife diversity as older stands, and plantations do not typically act as repositories for the same plant and wildlife diversity as natural forests. Therefore, the transformation of 16 million ha (9 million deforested and 7 million reforested as engineered plantations) necessarily altered the variety, quantity, and the pattern of delivery of plant and wildlife habitat that had been provided previously by mature natural forests. Biodiversity surely suffered.

Although the incremental 14 million ha of engineered stands supports its own increment of environmental services, those services will be very different from the environmental services provided by the forests that were standing in 1980. On a positive note, the 9 million ha increase in shelterbelts contributed in a beneficial way to the control of wind erosion, one of China's most pressing environmental problems. The net effect from the loss of biodiversity and the gain in erosion control is uncertain. However, the nature of China's wind erosion problem and the global uniqueness of its natural environment mean that both the favorable and problematic sides of the transformation of China's forests are critical environmental issues.

Standing Forest Volume

China's total standing forest volume remained virtually unchanged between 1980 and 1988, then rose sharply between 1988 and 1993 (Table 6-1). However, the distribution of this stock within each of the four major forest regions reveals a variable pattern of regional and intertemporal development. The pattern of aggregate growth was driven by the Northeast State Forest Region and the Southern Collective Forest Region, the traditional regions of industrial forestry. Standing volume actually declined in these two regions during the first period and recovered during the second (Table 6-1, Table 6-6). A pattern of regular growth in the small but robust North Central Fast-Growth Region partially offset a downward trend in standing volume for the Southwest Mixed-Management Region throughout both periods. (See Chapter 3 for an explanation of the relationships among reform policies, forest area, and harvest volume in the north-central region and the Southern Collective Forest Region in the early 1980s.)

At the provincial level, the volume of standing forests declined in the early period in all 10 of the most forested provinces except Jilin in the Northeast. Standing volume recovered in the later period in all provinces except Sichuan in the Southwest. The decline during the first period is typical of the resource mining that often occurs during the early stages of development. China's reforms seem to have halted this decline and prompted the reforestation that showed up as measurable increases in the standing forest stock by the late 1980s.

The increase in the standing stock in the later period reflects both new growth that occurred on the area of earlier increase in forest cover and a declining harvest rate that permitted more new growth to remain in the forest. Commercial timber harvests rose by 10% in the earlier years of reform but rose only 3% more between 1988 and 1993 (Table 6-6). This amounts to an annual rate of increase that is less than 0.6%, a remarkably low rate for an economy that was growing as rapidly as China's.

Why did roundwood production remain so remarkably steady for so long—at approximately 60,000 m^3/year? Perhaps it was due to the combination of strict government limitations on the use of wood as a building material (Richardson 1990), a restrictive permit system for the transportation of unprocessed wood products and high taxes on harvests from collective forests (Chapter 4), and rapid increases in the imports of paper and unprocessed wood products (*China Statistical Yearbook* 1989, 1994). The actual effects of these factors on production are unclear because the limitations on harvesting and transportation could not halt illegal logging, timber harvesting for local domestic uses or for the use of some traditional industries, or land conversion from forests to agricultural crops that promised higher returns (Zuo 1995, Qiao 1997). The official forest inventory was not designed to detect these factors.

Nevertheless, China's forest surveys show that its forests grew in terms of both forest cover and standing volume. Because roundwood production remained relatively stable between the second and third surveys, new sources must have been the basis for the large increase in standing volume. The obvious source is juvenile

Table 6-6. *Timber Resources and Production, Late 1970s to Early 1990s*

Region	Timber forest volume					Roundwood production				
	Million m³			% Change		Million m³			% Change	
	1980	*1988*	*1993*	*1980–1988*	*1988–1993*	*1980*	*1988*	*1993*[b]	*1980–1988*	*1988–1993*
Northeast State Forest Region										
Heilongjiang	1,341	1,216	1,267	–9.3	4.1	16	19	12	15.6	–34.9
Nei Monggol	797	775	798	–2.8	2.9	0.4	6	5	43.4	–14.8
Jilin	571	590	619	3.2	5.0	6	6	5	0.9	–15.8
Subtotal	2,709	2,581	2,684	–4.7	4.0	0.27	31	0.2	16.5	–27.2
Southern Collective Forest Region										
Fujian	284	245	286	–13.7	16.9	4	6	6	45.4	0.5
Jiangxi	229	157	167	–31.5	6.4	3	2	3	–21.5	11.0
Hunan	158	123	135	–21.7	9.6	2	3	3	–22.3	12.7
Guangdong[a]	194	162	187	–16.4	14.9	3	2	3	–30.8	38.1
Guangxi	174	156	154	–10.6	–1.4	2	2	3	30.7	37.6
Subtotal	1,039	843	929	–18.8	10.1	14	15	0.18	7.8	15.7
Southwest Mixed-Management Region										
Yunnan	976	698	662	–28.5	–5.2	2	3	4	35.0	9.9
Sichuan	767	713	737	–7.0	3.3	4	5	5	8.4	1.6
Subtotal	1,743	1,411	1,399	–19.0	–0.9	6	8	9	18.3	5.1
North Central Fast-Growth Region										
Jiangsu	2	4	5	84.8	15.3	0	0.1	0.5	—	939
Anhui	54	52	58	–3.6	10.0	0.5	0.4	2	–10	491
Shandong	2	6	7	238.6	35.4	0	0.1	1	119	1,772
Henan	19	24	30	27.5	24.4	0.2	0.2	2	3	1,031
Subtotal	77	86	100	11.7	15.9	0.7	0.8	6	8	790
All China	6,881.86	6,173.17	6,743.39	–10.3	9.2	54	62	64	16	3

Notes: The years 1980, 1988, and 1993 are approximate for the three forest survey periods.

[a]Hainan is included with Guangdong.

[b]For the North Central provinces, this estimate is based on mean timber production after they attained the new higher level of output in the period 1990–1992. For the others, it is based on the mean output over the entire period 1988–1993.

Sources: National Forest Resource Statistics 1983, 1989, 1994; *China Agricultural Yearbook* 1981, 1989, 1994.

growth on lands that were reforested or afforested for various purposes and with various species, mostly between 1980 and 1988 but also in earlier periods [3.3 million ha of fast-growing trees were planted in 12 provinces between 1974 and 1978 (CFFA 1979)]. Many of the newly reforested stands were too young to support measurable volume by the 1988 forest survey, but some of them were in their most rapid stage of young growth by the 1993 survey.

Multivariate Analytical Framework

We have presented a mostly optimistic assessment of the current state of China's forests. Forest cover expanded steadily, and the rapid decline in standing forest reserves that characterized the earlier reform period between the first two forest surveys appears to have been reversed most emphatically by the third forest survey in 1993.

To this point, our discussion has focused on understanding regional and intertemporal trends in forest cover and standing volume and has not developed our interest in the causes of these trends. In this section, we will design a simple empirical model to identify the causal factors for changes in forest cover and standing volume. We will consider the differential impacts of geography, climate, markets, and policy as well as the institutional distinctions between state and collective management.

Our general model takes the form

$$Y = \alpha_0 + \alpha_1 X_1 + \alpha_2 X_2 + \alpha_3 X_3 + \epsilon \qquad (6\text{-}1)$$

where Y is either the proportion of tree cover or the standing forest volume, or the change in one of these measures for the period between two forest surveys. The Xs are vectors. X_1 contains a set of economic factors likely to affect the supply or demand for trees, their products, and environmental services; X_2 contains measures of policy distinctions that are likely to affect forest management or timber removal; and X_3 contains regional dummy variables. The α_i are parameters to be estimated, and ϵ is a random error term.

Economic Factors

The endowments of labor and land affect forest production as well as the pressure put on the forest. We use rural population density to reflect labor availability and human pressure on the resource. Rural population density varied widely across China's regions, from 27.6 people per km^2 in the Northeast to 437 people per km^2 in the North Central Region. The pace of increase in this variable also ranged widely from negative in the Northeast to an annual increase of more than 0.5% in the Southern Collective Forest Region (Table 6-7a).

Land quality affects the area available for forestry because farmers with better agricultural lands are less likely to exploit their forests (Otsuka 1997). On the other hand, better quality land is more likely to be converted from forest to agricultural uses. We use the multiple cropping index as a measure of land quality.

This index varies widely between regions and over time. The smallest change occurred in the Northeast and the most rapid increase occurred in the Southwest (Table 6-7a).

General economic growth is likely to affect the various demands for marketed forest resources and resource services. It also is correlated with general welfare and external employment opportunities, forces that encourage environmental protection (welfare improvements) and take pressure off forest resources (employment opportunities). The share of light industrial production in total production is our measure of growth. It includes the construction and pulp and paper industries, the two largest sources of demand for wood fiber. The share of light industrial production increases with general economic growth for most economies, and it increased throughout the reform period for all but China's Northeastern State Forest Region (Table 6-7a). Once more, the Southern Collective Forest Region demonstrates the most dynamic growth. The share of light industrial production in this region increased at an annual rate of more than 1.5%.

The relative price of forest products also affects the forest resource. Our measure of this is the price of delivered roundwood relative to the price of grain. This relative price rose sharply throughout the early and mid-1980s, then declined slightly as agricultural prices rose more rapidly than roundwood prices in the mid-1990s. The effect of the relative price measure is uncertain, however, because increases in the relative price induce better forest management and expansion in the area of managed forest but they also encourage expansion in the harvested area and could lead to a contraction in the area of unmanaged natural forest. Conversely, decreases in the relative roundwood price reduce the level of forest management and increase agricultural land conversion from the formerly managed forest area, but they also cause a decline in harvests from the natural forest and recovery and expansion in the area of natural forest.

Policy Factors

Huang et al. (1998) and other chapters in this book have reviewed the changes in the forest policy environment during the reform period. Considerable doubt has been expressed regarding whether specific forest policy has hindered or helped forest resource restore the forest resource (Ross 1988, Richardson 1990, Smil 1993, Rozelle et al. 1998). In fact, in Chapter 3, Yin demonstrates that both effects were important, depending on the specific forest policy and its local application. We attempt to capture the overall policy impact with time dummy variables for 1988 (representing earlier policy changes) and 1993 (representing later policy changes). The 1988 policy dummy may be especially relevant as a summary indicator for the important policy changes that occurred early in the experience of China's overall reforms.[6]

Many have noted the important effects of land tenure on forestland use in other developing countries (e.g., Otsuka 1997, Hyde et al. 2000) as well as in

China (Chapters 2 and 4, Huang et al. 1998, Rozelle et al. 1998). In China, the important changes in tenure occurred in the southern, eastern, and north central regions, where villages and households obtained greater responsibilities and greater autonomy in their land-use decisions. The policy dummies will have positive coefficients if these effects were both immediate and relatively strong.On the other hand, some believe that the policies of the collectives subsequent to the initial reforms have failed (e.g., Qiao 1997). The policy dummies will be negative if this expectation is accurate.

The state forest farms managed large shares of the collectives' forests prior to reform. After 1981, these lands were increasingly returned to the collectives. Table 6-7b captures this redistribution. Managers of state-owned forest farms, primarily in the Northeast, were encouraged to reforest and to allow prices to reflect more accurately the true opportunity cost of timber. Nevertheless, annual production targets remained the primary incentives of these managers, even after reform. The failure to institute stronger incentives has probably made the state forest farms less committed to efficient operation. Therefore, we anticipate that the larger the proportion of collective forests managed by state-owned forest farms, the weaker the incentives for efficient management. A smaller proportion implies more land in collective and household management, where the incentives of improved tenure were greater.

Regional Distinctions

Our fundamental observations are for the 25 most forested provinces, but we apply dummy variables associated with China's seven traditional geopolitical regions to control for similarities in topography and climate. Those seven regions are: North China (*huabei*), Northeast China (*dongbei*), East China (*huadong*), Central China (*huazhong*), South China (*zhongnan*), Southwest China (*xinan*), and Northwest China (*xibei*). The Northwest is the least forested. Eliminating its dummy implies that the coefficients on the remaining dummies refer to a difference in impact from the control level of impact in the Northwest. In addition, we combine South and Central China, because the provinces of these two regions fall within the Southern Collective Forest Region. Four of the remaining five regions match the four forest management regions of our discussion to this point. It is more difficult to assign the fifth region (East China) to a specific forest management category.

Data

The fundamental forest data are from the three forest surveys described above. The remaining data on relative prices, population density, multiple cropping index, light industry share of gross domestic product (GDP), and tenure share are taken from the government's statistical publications (*China Agricultural Yearbook, China Statistical Yearbook*). In total, we have 25 provincial observations associated

Table 6-7a. *Factors Affecting Forest Management and Conservation, Late 1970s to Early 1990s*

	Rural population density				*Multicropping Index*			
	÷ 10,000 People/km²			*% Change*	*Index*			*% Change*
Region	*1980*	*1988*	*1993*	*1980–1993*	*1980*	*1988*	*1993*	*1980–1993*
Northeast State Forest Region	0.00291	0.0027	0.0028	–5.1	98	95	98	0.8
Southern Collective Forest Region	0.00194	0.0020	0.0021	7.8	210	212	223	6.1
Southwest Mixed-Management Region	0.00126	0.0013	0.0013	5.6	170	179	191	12.7
North Central Fast-Growth Region	0.00418	0.0042	0.0044	4.6	161	173	174	7.5
All China	0.0089	0.0090	0.0094	5.6	147	151	155	5.4

Notes: Values for 1980, 1988, and 1993 are approximate for the three census periods.
Source: China Agricultural Yearbook 1981, 1989, 1994.

Table 6-7b. *Factors Affecting Forest Management and Conservation, Late 1970s to Early 1990s*

	Total value of output from light industry				*Collective forest area managed by state-owned forest farms*			
	Share			*% Change*	*Share (%)*			*% Change*
Region	*1980*	*1988*	*1993*	*1980–1993*	*1980*[a]	*1988*	*1993*	*1980–1993*
Northeast State Forest Region	0.265	0.287	0.255	–3.6	100	91.9	93.2	–6.8
Southern Collective Forest Region	0.347	0.393	0.434	25.1	100	13.9	13.1	–86.9
Southwest Mixed-Management Region	0.295	0.318	0.317	7.5	100	43.9	41.8	–58.2
North Central Fast-Growth Region	0.362	0.386	0.397	9.7	100	13.5	14.7	–85.3
All China	0.355	0.373	0.379	6.8	100	51.5	51.9	–48.1

Notes: Values for 1980, 1988, and 1993 are approximate for the three census periods.
[a]Assumed to have been 100% under management of the state-owned forest farms through 1980.
Sources: For total value of output from light industry, *China Statistical Yearbook* 1980 through 1994; for collective forest-area managed by state-owned forest farms, *National Forest Resource Statistics* 1983, 1989, 1994.

with the time periods of the three forest inventories, or 75 observations for the regressions estimating the determinants of levels of tree cover and standing forest volume and 50 observations for the regressions estimating the determinants of changes in the levels of tree cover and standing forest volume.

Results

Table 6-8 shows the ordinary least squares regression results for two sets of two equations, one set evaluating effects on forest cover and the other evaluating effects on standing forest volume. Each set shows the coefficients explaining effects on the level and also on the change in the dependent variable. The equations performed well in general. The measures of goodness of fit (adjusted R^2) range from 0.45 to 0.87 for the forest cover equations and from 0.19 to 0.71 for the volume equations. The R^2 values are lower for the change equations, as expected. Most coefficients have reasonable signs, and the coefficients are robust with respect to various adjustments in equation specifications.

Determinants of Forest Cover

Rural population density has a negative and significant effect on the level of forest cover and a nearly significant effect on changes in forest cover. Greater population density means more labor. When labor is abundant, the wage is low, and low-wage workers rely heavily on the natural forest and harvest farther into its interior. Forest cover declines as a result (Foster et al. 1997, Hyde et al. 2000). A 1% increase in rural population density caused a 0.59% decrease in forest cover.

The positive coefficients on the multiple cropping index indicate that better agricultural land quality is associated with more intensive agricultural production and less forest conversion, although the small and insignificant coefficient in the change equation means that agricultural land quality made little difference to forest cover during the reform period. This finding indicates complementarity between agricultural and forest investments. While it is fairly novel in the general development and resource literature, this finding is consistent with the observations of Huang et al. (1998) at the village level. It means that when the government makes public investments that improve farmland or when farmers improve farmland in response to investment incentives, the investments help increase forest area as well.

The large and significant coefficients on the light industry term indicate that general economic growth had a positive effect on forest cover. A 1% increase in the light industry share of GDP caused a 0.13% increase in forest cover. Three factors could be at work. Industrial demands for wood products could have induced planting and net gains in forest cover (i.e., more new plantations than deforestation of natural stands). The stronger industrial demand also could have had an indirect effect if increasing industrial wages attracted labor to the industrial sector and away from forest-dependent activities. Finally, there could have

Table 6-8. *Regression Results for Forest Cover and Volume, Levels and Changes, 1980–1993*

	Dependent variable			
	Forest cover		Standing forest volume	
Independent variable	*Level*	*Change*	*Level*	*Change*
Constant	−14.45	—	−1,939.40	—
	(1.66)		(1.87)	
Rural population density	−526.60	−58.56	−46,260	0.32
	(5.92)	(1.42)	(4.38)	(1.06)
Land quality (MCI)	0.22	0.0001	2.62	0.000002
	(7.72)	(0.20)	(0.76)	(0.66)
Share of GDP from light industry	41.73	13.26	4,636.20	0.04
	(3.60)	(3.61)	(3.37)	(1.45)
Relative price index (delivered logs/grain)	−8.79	−0.55	241.25	−0.002
	(1.28)	(0.49)	(0.30)	(0.27)
Tenure (share of collective forest area under state management)	−1.84	−1.40	724.43	−0.007
	(0.52)	(0.94)	(1.72)	(0.62)
Policy effect (year dummy—1988)	5.46	−1.01	180.10	0.0005
	(0.89)	(0.91)	(0.25)	(0.06)
Policy effect (year dummy—1993)	6.78	—	249.12	—
	(1.02)		(0.31)	
Regional dummies				
Northeast	27.15	−0.02	2,258.34	−0.001
	(12.29)	(0.01)	(8.61)	(0.09)
North	5.48	0.99	389.78	−0.003
	(2.27)	(0.96)	(1.36)	(0.34)
East	3.76	2.36	1,172.24	−0.02
	(0.89)	(1.60)	(2.35)	(1.69)
Central/South	0.78	0.15	834.25	−0.01
	(0.21)	(0.12)	(1.87)	(1.60)
Southwest	−1.38	1.21	1,750.01	−0.007
	(0.46)	(0.88)	(4.91)	(0.66)
Adj. R^2	0.87	0.45	0.71	0.19
Observations	75	50	75	50

Notes: Ordinary least squares estimations; *t*-ratios in parentheses. MCI, multiple cropping index; GDP, gross domestic product.

been a wealth effect: as the industrial sector and the entire economy grew relative to the forestry sector, the entire economy's greater wealth could have been associated with an increasing demand for the nonconsumptive conservation services of forests and, therefore, with an increase in forest cover—or at least a slowing of the rate of destruction of the natural forest. In addition, wealthier households tend to prefer coal and natural gas rather than wood as a source of heat and light, and this difference would have decreased the demand on the forest as a source of fuel.

The relative price variable was not significant in any of our equations. Its insignificance may mean that the opposing effects of relative prices approximately offset each other. In other words, higher relative roundwood prices may have induced approximately comparable reductions in natural forests (due to the incentive to increase the harvest level) and additions to new managed forests (due to the incentive to plant in order to satisfy a higher expected level of future demand).

The negative sign on the tenure variable indicates that forest cover decreased as the share of all collective forests managed by state-owned forest farms increased. Provinces with larger concentrations of collective lands that are managed by state forest farms have lower levels of forest cover and have experienced sharper declines in forest cover since the beginning of reforms. This finding is important for those who have lamented the poor performance of the collective sector. It indicates that the collectives and individual households have done better than the state-owned forest enterprises in increasing forest cover. In fact, the tenure variable has the most elastic effect of any of our variables on forest cover. A 1% change in the lands that shift from state to collective or household management causes a 1.4% increase in forest cover. This elasticity demonstrates the importance of providing better incentives in forest management.

The time dummy variables generally have little predictive power. The regional dummy variables show that all regions (except perhaps the Southwest) had the expected higher levels of forest cover than the arid Northwest. If these time dummies are proxies for forest policy, they would mean that forest policies, apart from those that affect tenure and prices, have been ineffective. These same regions (with the exception of the Northeast State Forest Region) had higher levels of growth in forest cover as well.

In sum, the critical factors for explaining changes in forest cover are population (or labor) and general economic activity (the light industry variable). General economic activity had a strong positive influence. That is, forest policy may have been important, but the performance of the overall economy was more significant. This finding is especially interesting because general economic performance is commonly overlooked when analysts look for factors that improve the performance of the forestry sector.

Determinants of Standing Forest Volume

The signs and significance of the volume regressions are largely consistent with those observed for the forest cover regressions, although the elasticities of the effects were much smaller because the changes in standing forest volume were smaller than the changes in forest cover. This consistency with the volume regressions is especially true for the economic determinants (population density, general economic development, and relative price). Wealthier regions and regions with lower levels of rural population pressure contributed positively to the increasing levels of forest volume across China over time. Land quality also may have had a positive effect. Higher relative prices for roundwood had offsetting effects, increasing the volume harvested but encouraging planting and greater volumes of new growth.

The large positive and significant coefficient on state tenure in the level equation reflects the initial policy bias that gave state forest farms control of the best timber resources (Rozelle et al. 1998). Therefore, land redistribution from the state to the collectives may have contributed to an increase forest volume. This change in tenure had the most elastic effect of any variable on forest volume (e = 0.007), but the effect was statistically insignificant.

The time dummy variables indicate that general policy reform may have led to growth in forest volume. However, neither these time dummies nor those for the forest cover equations are statistically significant. Therefore, we cannot draw any convincing conclusions about overall policy effects in either period.

The regional dummy variables show that several regions contained larger initial volumes than the control region, the dry and relatively less forested Northwest. Growth in all regions represented by dummy variables was less than new volume growth in the Northwest. This difference could be explained by the young growth in these regions that began to show up in the measures of forest inventory only toward the end of the period of our analysis.

Summary and Conclusions

The data from the three forest inventories preceding and immediately following the implementation of rural reforms provided a detailed and disaggregated picture of China's forests. The data provide good reason for qualified optimism: a 15% increase in aggregate forest cover from 1980 to 1993, and a sharp recovery in forest volume in the past decade. These increases, from a mix of timber tracts, shelterbelts, and forest plantations, turned around the decline in forest cover and volume that had been occurring since the beginning of the communist state, or even before.

The nature of the reforms shifted between the earlier and later periods in terms of its emphasis on the form and administration of forest management. The magnitudes of investment in the agricultural and nonagricultural sectors and several key economic factors also changed. The economic and policy factors varied across China's regions, and the rates of forest growth varied across regions as a result. For example, the percentage change in tree cover was much greater in the later period (1988–1993) than in the earlier period (1980–1988) for both the predominantly state-run Northeast and the predominantly collective-run South. In the newly important North Central Fast-Growth Region, the trends of several factors differed from other regions, and the differences may account for the reversal of the prior pattern of change in regional resources. The rapid growth of agricultural investment in this region early in the reform period may account for the early expansion in this region's forest resources. In the Southwest Mixed-Management Region, part of the increase in forest resources was clearly due to large centrally directed investments in shelterbelts in the upper and middle reaches of the Yangtze River Valley.

Our optimism must be guarded, however, because part of the increase in forest cover has apparently come at the expense of forest diversity. During the first 15 years of reform, China transformed the structure of at least 6% of its forest area, changing natural forests into plantations. Although such a transformation may increase individual profits and lead to improved delivery of environmental services such as erosion control, it does not necessarily improve the ability of this 6%—or the additional 15% of newly forested area—to provide environmental services such as habitat for threatened and endangered species. The change in forest structure also may affect the livelihoods of certain groups of rural peoples who rely on products that were a component of the natural forest but are not available in the new managed forests (e.g., certain herbs, medicinal plants, and forage and fodder for domestic livestock). These issues are clearly important, and they deserve careful attention. The nature of our data set, unfortunately, does not provide enough information to pursue this point.

So what must be done, policywise? Several of our findings provide clear signals regarding the direction of future policy if the nation's leaders want to foster the current pattern of forest development, a pattern that certainly is consistent with the nation's next 5- and 10-year plans (*China Agricultural Yearbook* 1995 through 1997). In one sense, China's leaders should continue doing what they are doing, deepening current reforms and allowing newly emerging economic activities to continue to develop. Our assessment has shown that continued growth in the economy and continued improvements in the incentives faced by all forest producers—the leaders of the collectives, individual villagers, and especially the managers of the state-owned forest farms—will encourage investment in new forests and plantations. The national plan to increase agricultural investments, detailed in a Ministry of Agriculture white paper (MOA 1997), should have a positive impact on forest resources, as should continuing trends for farmers to move out of the agricultural sector into the off-farm job market (Rozelle et al. 1998).

Serious challenges still face China's forestry leaders despite their past successes and guarded optimism about the future. Our accounting of the changes in the various categories of forest points to the ineffectiveness of current forest policies and administration for monitoring and resource extraction—that is, the natural forest continues to decline in area and volume. Furthermore, although rising wealth appears to be a source of improvement in the overall level of forest resources, increasing income inequality and patterns of persistent poverty may spell trouble for forest resources in areas that have not been integrated into China's drive for rapid economic growth. Additional research is needed on these elements and on the ways to encourage forest users to expand the area and volume of the resource without destroying its diversity.

Notes

1. The Ministry of Forestry was broken into three components: the part that manages nature reserves and national parks was moved to the state environmental protection agency; the part that manages most major afforestation, reforestation, and new shelterbelt activities was moved directly under the State Council; and the remainder of the ministry and its logging and manufacturing enterprises are being commercialized.

2. Fewer resources have been devoted to and less monitoring has been conducted in the collection of China's forest statistics than in the collection of its agricultural statistics. There may be a number of reasons: the authorities place lower priority on forest products relative to grain and other agricultural crops, forest products often originate from poorer and more remote areas where monitoring is more difficult, and a data-quality problem may arise from the dual nature of the sector. That is, information on one-half of the sector's output is reported through the agricultural and forestry reporting system (which is coordinated by the statistical bureau), whereas the other half is collected by enterprises in the state-owned industrial sector and reported through a different channel.

3. From interviews with officials at the State Environmental Protection Agency and our own interviews in a nearly national survey of 220 representative villages across China (Rozelle, unpublished notes).

4. In fact, the net adjustment was even greater by the unidentified level of "unplanned" tree planting by farmers on land that had no previous forest cover.

5. This observation is consistent with a previous analysis of the changes in the age structure of China's forests (Rozelle et al. 1998). Young seedlings do not contain volume, as it is measured in most forest inventories. Trees begin to meet the minimum standards for volume in China's inventory after three to five years. Therefore, trees planted within two years of the next inventory are accounted for in estimates of forest cover but not in estimates of standing forest volume.

6. The regressions examining change in forest cover or change in standing volume drop one time dummy because the data permit only two periods of change and one period must act as a control. Our regressions retain the 1988 dummy on the grounds that the package of earlier reforms is more widely discussed and may have had greater impact. It is more interesting to examine.

References

CFFA (*China Facts and Figures Annual*). 1979. Gulf Breeze, FL: Academic International Press.

China Agricultural Yearbook. 1981, 1989, 1994, 1995, 1996, and 1997. [In Chinese]. Beijing: China Agricultural Press.

China Forestry Yearbook. 1986. *China Forestry Yearbook, 1949–1986* [in Chinese]. Compendium volume. Beijing: China Forestry Press.

China Forestry Yearbook. 1987 through 1993. [In Chinese]. Beijing: China Forestry Press.

China Statistical Yearbook. 1980 through 1994. [In Chinese]. Beijing: China Statistical Press.

Foster, A., M. Rosenzweig, and J. Behrman. 1997. Population and Deforestation: Management of Village Common Land in India. Draft manuscript. Philadelphia: University of Pennsylvania, Department of Economics.

Ghimire, Krishna. 1994. Conservation and Social Development: A Study Based on the Assessment of Wolong and Other Panda Reserves in China. Research working paper. Geneva, Switzerland: United Nations Institute for Social Development.

Harkness, J. Forthcoming. Forest and Conservation of Biodiversity in China: Threats and Prospects. *The China Quarterly.*

He, B. 1991. *China on the Edge: The Crisis of Ecology and Development.* San Francisco: China Books & Periodicals, Inc.

Huang, Jikun, Scott Rozelle, and Fangbin Qiao. 1998. Private Holdings, Conservation, and the Success of China's Forest Policy in the Reform Era. Working paper. Beijing: Chinese Academy of Sciences, Center for Chinese Agricultural Policy.

Hyde, William F., Gregory S. Amacher, and colleagues. 2000. *The Economics of Forestry and Rural Development: An Empirical Introduction from Asia.* Ann Arbor: University of Michigan Press, Chapter 10.

Jefferson, G.H., and T.G. Rawski. 1994. Enterprise Reform in Chinese Industry. *The Journal of Economic Perspectives* 8(2): 47–70.

Lin, Justin Yifu. 1992. Rural Reforms and Agricultural Growth in China. *American Economic Review* 82(1): 34–51.

McMillan, John, and Barry Naughton. 1992. How To Reform a Planned Economy: Lessons from China. *Oxford Review of Economic Policy* 8: 130–143.

MOA (Ministry of Agriculture). 1997. *White Paper on Agricultural Policy in 1996* [in Chinese]. Beijing: China Agricultural Press.

National Forest Resource Statistics. 1983 through 1989, 1991, 1994. [In Chinese]. Beijing: Ministry of Forestry.

Otsuka, Keijiro. 1997. Property Rights and Forests in Africa. Working paper. Washington, DC: International Policy Research Institute.

Qian, Yingyi. 1994. Lessons and Relevance of the Japanese Main Bank System for Financial System Reform in China. In *The Japanese Main Bank System: Its Relevancy for Developing and Transforming Economies,* edited by M. Aoki and H. Patrick. Oxford, U.K.: Oxford University Press.

Qian, Yingyi, and Chenggang Xu. 1993. Why China's Economic Reforms Differ: The M-Form Hierarchy and Entry/Expansion of the Non-State Sector. *The Economics of Transition* 1(2): 135–170.

Qiao, Fangbin. 1997. Property Rights and Forest Land Use in Southern China. Unpublished master's thesis. Beijing: Chinese Academy of Agricultural Sciences Graduate School.

Richardson, S.D. 1990. *Forests and Forestry in China.* Washington, DC: Island Press.

Ross, Lester. 1988. *Environmental Policy in China.* Bloomington: Indiana University Press.

Rozelle, Scott, Li Guo, Minggao Shen, Amelia Hughart, and John Giles. 1998. Poverty, Networks, Institutions, or Education: Testing Among Competing Hypotheses on the Determinants of Migration in China. Working paper. Davis: University of California–Davis, Department of Agricultural Economics.

Schur, Catherine. 1984. Hainan Dao: Contemporary Environmental Management and Development on China's Treasure Island. Ph.D. dissertation. Los Angeles: University of California–Los Angeles, Department of Geography.

Smil, Vaclav. 1984. *The Bad Earth: Environmental Degradation in China.* New York: M.E. Sharpe.

———. 1993. *China's Environmental Crisis: An Inquiry into the Limits of National Development.* New York: M.E. Sharpe.

Sun, Changjin. 1992. Community Forestry in Southern China. *Journal of Forestry* 90(6): 35–39.

World Bank. 1992. *China Environmental Strategy Paper.* Washington, DC: World Bank.

Yin, Runsheng. 1994. An Empirical Analysis of Rural Forestry Reform in China, 1978–1990. Ph.D. dissertation. Athens: University of Georgia, Warnell School of Forest Resources.

Zuo, Ting. 1995. Nontimber Forest Use in China. Working paper. Kunming, China: Yunnan Provincial Academy of Social Sciences, Rural Development Institute.

CHAPTER 7

Deforestation and Reforestation in Hainan

Roles of Markets and Institutions

Yaoqi Zhang, Jussi Uusivuori, Jari Kuuluvainen, and Shashi Kant

THIS CHAPTER CONTINUES the examination of the sources of change in China's forest cover since 1978. Rozelle, Huang, and Benziger examined this question in Chapter 6 with data for all of China. They concluded that the changes in land tenure that have been a focus of discussions of agricultural and forest reforms may have had an important role. However, they observed that general economywide growth had a more significant effect on the forestry sector than any particular forest policy reform. This observation indicates that macroeconomic policy and growth can be more important than fine-tuned policies that are specifically designed for small sectors such as forestry. Since our interests in forests reflect our environmental concerns as well as our concern for the sector's market production, these observations also remind us of the important effect of aggregate economic welfare on the condition of selective natural environments.

Rozelle et al. reminded us of the distinction between managed forest plantations and natural stands, a distinction that is important for the provision of environmental services and for understanding the forests' response to market signals. However, they did not consider that distinction in their econometric analysis. We will emphasize that distinction as we conduct a similar search for the causal factors of forest development in China since the beginning of reforms in 1978. Our focus on the island province of Hainan, however, will be narrower than the all-China perspective of Chapter 6. The narrower geographic focus has its advantages. Because Hainan is an island, its market boundaries are clearer, and its smaller size means that there are fewer institutional variations in policy applications than some previous chapters have noted for the whole of China. Therefore,

---

This paper is one component of Yaoqi Zhang's Ph.D. thesis at the University of Helsinki, Finland. A prior version appeared as Zhang et al. 2000. The Economy and Environment Programme for Southeast Asia provided financial assistance.

we anticipate that our problem and our data may be more sharply defined and less subject to unknown external influences—and that our results may be more conclusive.

We will begin with a discussion of Hainan and then turn to our econometric model and its results.

Forestry and Economic Reform in Hainan

Hainan is a tropical island of 3.4 million ha in the South China Sea. It is almost as large as Taiwan. Hainan was upgraded to a province and designated a Special Economic Zone in 1988. Since then, its population has grown from 5 million to 8–10 million, and its economic growth has averaged 20% annually—the fastest of any province in China.

Forest Cover before Reforms

Hainan was originally covered by tropical rainforest, and rainforests still covered 50% of the island in the early 1930s, at the beginning of the Japanese occupation. Four Japanese firms with 1,500 loggers extracted more that 10,000 m^3 annually to support 18 new wood-processing establishments (Chen 1948). Subsequently, during World War II, the Japanese removed an additional large amount of timber for road construction and war preparation, and still more was destroyed by forest fire.

By the early 1950s, Hainan's forest cover had declined to about 30% of its total land area. Most of the remaining forest was located in the central mountain ranges (HAZC 1980). Forest cover continued to decline in response to expanding demands for wood from within the island and also from external markets. About 40,000–60,000 m^3 of tropical logs flowed off the island annually from the late 1950s to the early 1970s. Forty thousand hectares of rainforest had been logged by the end of the 1980s, with very little silvicultural reinvestment. After repeated high-grading, much of Hainan's former forest became degraded shrubland or even bare land, and Hainan went from being a net exporter of timber in the 1950s to an importer in the 1980s.

Tropical tree crops (e.g., rubber, tea, and tropical fruits) were introduced at the beginning of the twentieth century, but the land area in these crops did not change significantly until the 1950s, when it expanded dramatically (Figure 7-1). China's increasing domestic demand for natural rubber was a critical factor because trade with the West was blocked at the time. The central government organized a large-scale Farm Reclamation System (FRS) to plant rubber on 0.85 million ha, approximately one-fourth of the island's total land area. The FRS also administered 450,000 ha of natural forest, which declined to less than 100,000 ha by 1980 (Situ 1992).

Agricultural land use did not expand significantly over this period, but shifting cultivation—widely practiced by the local minority population—destroyed large

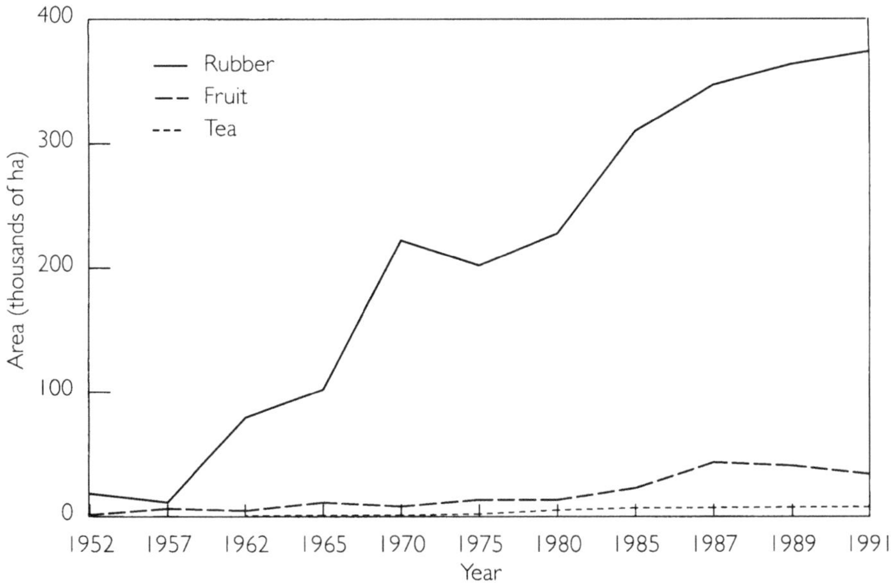

Figure 7-1. *Tropical Tree Crop Expansion in Hainan, 1952–1991*

tracts of rainforest. Bao (1991) estimated that shifting cultivation destroyed 2,700 ha of forest annually in the 1970s and 1980s. By the late 1970s, Hainan's forests had shrunk to their lowest level, 15% of total land area, and degraded lands had become the largest category of all forestland, accounting for one-fourth of Hainan's land area (HAZC 1980).

China began its gradual movement toward a market economy in the late 1970s. The international market for eucalyptus woodchips was expanding at that time, and Hainan was geographically positioned to take advantage of it. The Ministry of Forestry (the State Forest Administration since 1998) began to establish 130,000 ha of fast-growing plantations of eucalyptus and a few other species in 1982. The ministry provided the capital, local collectives provided the land, and local people provided the labor. Several joint plantation projects followed, some with financial assistance from international lending authorities like the World Bank. In general, these projects were successful. They accounted for more than one-third of all plantation forests by 1995 and contributed to a 4% increase in Hainan's forest cover. Once these plantations were mature, they became an important source of foreign currency. About one million tons of wood chips were exported between 1989 and 2000.

In sum, the long trend of declining forest cover reversed in the late 1970s, and the land area in forests expanded through the 1990s (Figures 7-2 and 7-3). Therefore, Hainan provides empirical support for the "forest transition" hypothesis, which suggests that forest cover first declines with initial economic growth and then eventually begins to increase with further growth (Mather 1990, Hyde et al. 1996, Rudel 1998, Zhang 2000).

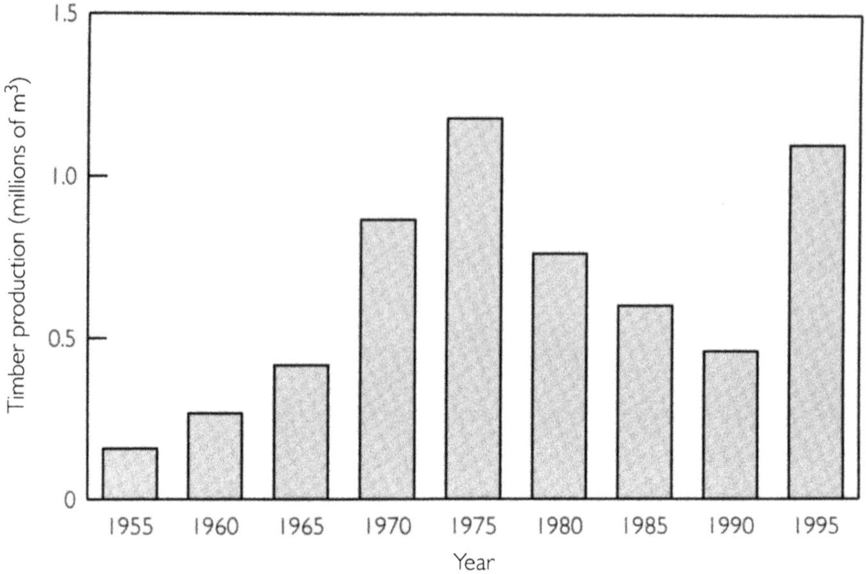

Figure 7-2. *Timber Production, 1950s–1990s*

Policy

China's land reforms of 1950–1956 confiscated all land owned by landlords and redistributed most of it to local farmers. Land in remote areas, particularly the rainforests in the central part of Hainan, was allocated to state-owned logging firms. About 350,000 ha were allocated to these firms, 80% of it to the 11 largest firms (HAZC 1980).

Land was reallocated once more in the "socialist transformation" that began in 1957. It was transferred from individual farmers first to the collectives, then to the people's communes. The people's communes continued to exist until the death of Mao Tse-Tung in 1976, although there were a couple episodes of decollectivization [e.g., after the collapse of the Great Leap Forward in the late 1950s (Walker 1965)]. Compared with agriculture, the forestry sector remained much more collectivized and centralized until the beginning of economic reforms in 1978.

The household responsibility system (HRS) in agriculture highlighted the early economic reforms. HRS actually represents a kind of privatization of land-use or labor-use rights. Reforms in the forestry sector began a little later than in agriculture, and they have been both less intensive and less extensive in forestry than in agriculture (Zhang et al. 1999). The key features of forest policy reform were:

- Control of public forestland nominally owned by the state was transferred to state-owned enterprises.
- Management authority for state-owned forests was transferred from national authorities to provincial authorities, then to county authorities.

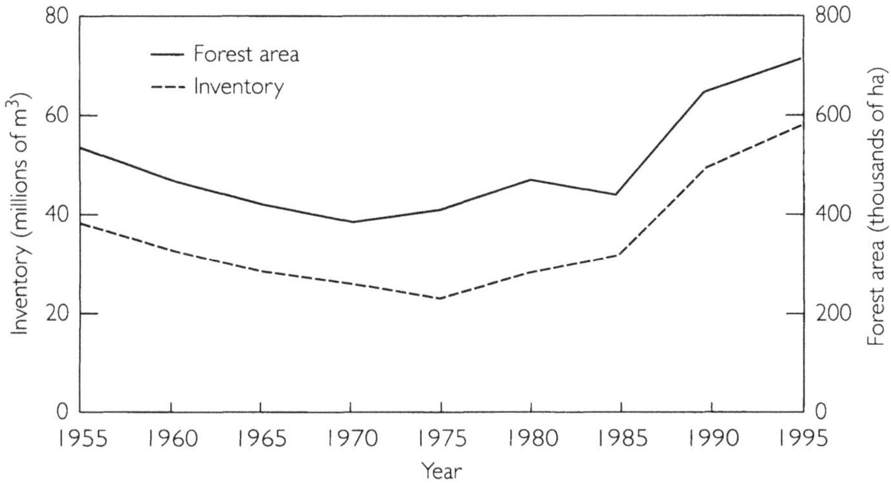

Figure 7-3. *Forest Cover, 1950s–1990s*

- Collective forestland was distributed to households. Alternatively, parts (or all) of the management responsibilities for this land were transferred from the collectives to contractors.
- Property rights were redefined by organizing various joint ventures for forest management that combine groups of state or community or private inputs.

The impact of HRS in agriculture was reflected in sharp increases in agricultural productivity and efficiency, and these reduced the pressure to convert forestland to agriculture. The greater impacts of HRS on forestry up to now lies less in increased production than in the clarification of property rights and the reduction of uncertainty. (Previous chapters have reviewed these reforms and the changes in forest cover and volume.)

Today, forest tenure and administration in Hainan closely resembles that in the rest of China. State-owned forests and collective forests coexist in Hainan, and Hainan provides a good opportunity to examine the contrasting effects of both systems of tenure and administration within one province. (One system or the other tends to dominate in most other provinces and regions.)

While the overall pattern of reform has been toward a more open market, this pattern has not been universal. Government agencies at various levels have imposed additional regulations on forest management from time to time. First, only selective cutting was permitted in the late 1970s. Then, a quota for allowable timber harvests was established in 1984 and gradually reduced since. Finally, for Hainan, all harvests from the remaining rainforests were prohibited in 1993. Meanwhile, many degraded areas were closed to accelerate their rehabilitation. One hundred thousand ha were rehabilitated in the 1980s, and a similar area was rehabilitated in the 1990s (Zeng 1994). In total, 6–8% of the island has been rehabilitated or reforested as a result of forest closures and contracts between the collectives and local households who guard the forest in return for an annual fee or a share in the final timber revenues.

The ban on all harvests from the rainforest reflects the growing concern for the environmental services provided by the natural forest: protection against soil erosion and the provision of water supplies, habitat for endangered species, and even opportunities for ecotourism. However, the state-owned forest bureaus and the local economy depended on logging, and they have suffered greatly as a result of the ban. The national and provincial governments have compensated for some of the loss, and they have encouraged low-interest loans to finance a shift to other economic activities, including silvicultural management outside the rainforest, furniture production, and ecotourism. A decade after the ban, the forests are recovering, even if the local economy still faces some difficulties.

A Formal Model of Land Allocation

Hainan's forest history has been a story of, first, deforestation of the natural rainforest, then reforestation and forest management. The modern managed forest includes both plantations and actively managed secondary natural forest. The managed forest also includes a small but important share of tropical tree crops such as rubber and fruit. Our primary interests are the first two classes of forest as the main sources of both industrial timber and many forest-based environmental services. These two classes are also the principal focuses of general policy interests in global forestry.

Although the same economic principles apply to both managed forests and rainforests, each responds differently to prices and to various institutional factors. The price responses of the managed forests resemble the responses observed in sectors like agriculture, whereas the price responses of the natural forest are similar to those observed in mining. We expect that rising product prices are an incentive to expand agriculture and managed forests. They are an incentive to deplete mines and the natural rainforest. Public forestry institutions initially encouraged logging in the natural forests. Only more recently have they begun to encourage reforestation in managed forests and protection of the natural forest.

More formally, land allocation to either category of forest A_i (where the subscript $i = 1, 2$ refers to the selection of either natural rainforests or managed stands as a dependant variable) is a function of output prices p, input costs c, and the total area of available land A. Optimal land allocation is nondeclining in own prices, and the cross-price effects are nonpositive. This means that land allocation to any particular use increases with its own price, and we expect that it decreases with the prices of competitive land uses. Changes in cost have an indeterminate effect that depends on the substitution between land and other inputs, plus an output effect.

In addition to the direct market effects, the broader economy E and institutional factors I play roles in land allocation. Therefore, we include variables describing the economic development of the region and the rights to forest properties in our model of land use. We expect that an increasing level of general economic activity creates a greater demand for industrial wood products. It increases the demands to deplete the natural rainforest but also increases the incentives to expand the managed forest. Improvements in the institution of property rights increase the efficiency of the system and improve the return on investment in forest management.

The final conceptual model is

$$A_i = A_i(p,c,A;E,I) \qquad (7\text{-}1)$$

If managed and natural forests respond differently to some critical independent, as we have argued, then we must examine the factors affecting these two classes of forest in separate regressions.

Data and Econometric Specification

Hainan is administered as 19 counties and one state reclamation bureau. Frequent changes occurred in the administrative boundaries of the counties, but we can accommodate these by combining some counties. We eliminated the state reclamation bureau because the data referring to it are sparse. This leaves 13 cross-sectional units and data for 17 years between 1957 and 1994. (Data are not available for every year.) The forest cover data are from the statistics of the forest bureaus maintained at the Institute of Forest Resources Investigation (Hainan Forestry Bureau 1957 through 1995). Only the interior counties contain significant amounts of natural forest, and their harvests have been strictly limited since 1985. Therefore, the regression identifying the determinants of rainforest cover are based on data from eight cross-sectional observations of those counties with more than 2% land cover in rainforest and for the period 1957–1985.

The price data refer to delivered logs, agricultural crops, and tropical tree crops. The delivered log prices reflect the principal products of the two classes of forests, eucalyptus for managed forests and tropical hardwoods for the natural forest. We separated tropical tree crops from agriculture because their price series display different trends. Furthermore, agriculture and tropical tree crops tend to use lands of different quality. Therefore, they compete with forestry in different ways. All four price series are composed of price indices divided by the index for rural industrial materials (following Yin and Newman 1997) and deflated to a 1957 base (Hainan Statistics Bureau 1957 through 1995).

We follow Rozelle et al. in Chapter 6 in using a measure of population density *PD* to reflect labor availability. Labor is the primary cost item in forest management and logging. Of course, increased population density could also indicate increasing pressure on the forest.

Gross output value per capita (*GOP*), deflated by the retail price index, is our measure of broader economic effects on the forests. Data on gross domestic product (GDP) were unavailable in the earlier statistical reports for the island, and *GOP* should be a reasonable proxy for GDP.

We discussed the changes in land-use rights that occurred, especially since 1978. The granting of rights to households under the household responsibility system improved the market incentives. The share of land under HRS contracts is a measure of this effect. The period 1980–1987 was characterized by uncertainty concerning the direction of economic reforms. Farmers and even local and provincial officials were uncertain how the reforms would be carried out. Their administration varied between jurisdictions, and some HRS contracts were even rescinded. Forests across the country suffered from this uncertainty. A dummy variable *U* captures the effect for these eight years. We introduce it in the regressions in a manner that allows both the slope and the intercept of decollectivization (*HRS*) to vary.

Finally, a substantial share of forestland remained the responsibility of the state-owned forest bureaus, and this share varies with the county administrative units. The share of state-owned lands (*SF*) is a measure of this institutional effect. The management incentives for these lands were tied to output quotas and employment guarantees, and the counties with larger shares of state-owned forests were not as responsive to market variables. Data on land shares are from the Hainan Agricultural Zoning Commission (HAZC 1980) and the *Hainan Statistical Yearbooks* (Hainan Statistics Bureau 1957 through 1995). Both measures of institutional change, *HRS* and *SF*, were determined by central government authority. Therefore, we can be treat them as exogenous (Besley 1995).

The final econometric specification is

$$FC_{jt} = \alpha_0 + \sum_{j=1}^{12} \alpha_j D_j + \sum_{n=1}^{3} \beta_n p_{n,jt} + \beta_4 PD_{jt} + \beta_5 GOP_{jt} + \beta_6 HRS_{jt} + \beta_7 U_t + \beta_8 HRS_{jt} U_t + \beta_9 SF_{jt} + \epsilon_{jt} \quad (7\text{-}2)$$

where *FC* represents forest cover in natural or managed stands as a share of total land area in each county and the subscripts *j* and *t* indicate the counties and years, respectively; the D_j are regional dummy variables; α_0 is a constant; the βs are parameters to estimate; and ϵ is a normally distributed random error. The regional dummy variables account for unobserved county-specific institutional and environmental factors. All variables except *D, HRS, U,* and *SF* are in natural logarithms. Therefore, β_1 through β_5 are elasticities.[1]

The two regressions were estimated according to the iterative generalized least squares method, with a Cochrane–Orcutt transformation to account for the relatively strong autocorrelation in the data (Kmenta 1986, Greene 1997).[2] Table 7-1 displays the results. (The cross-sectional dummies are likely to display contemporaneous correlation because macroeconomic, as well as geographic and socioeconomic, factors are likely to affect land allocation in all of the counties.) The regression results are most satisfactory. The log-likelihood is in a satisfactory range, and 16 of 18 coefficients are significant in the two equations.

Estimated Results: Levels of Forest Cover

Higher timber prices promote investment in managed forests, as expected. However, higher timber prices are also incentives to harvest the natural forest, and the land area in mature rainforest decreases as the timber price increases. The increasing hardwood price caused the expansion of mining-type harvest activity in the rainforest, at least until the ban on rainforest harvests in 1985.

Table 7-1. *Generalized Least Squares Results for Managed and Natural Rainforest Cover, 1957–1994*

Variable		Cover (%)	
		Managed forest	*Rainforest*
α_0	Constant	–3.74	2.98
		(7.03)	(3.40)
p_t	Price of timber	0.61	–1.52
		(5.81)	(4.80)
p_a	Price of agricultural products	0.11	0.33
		(1.16)	(6.55)
p_c	Price of tropical tree crops	–0.40	1.29
		(4.22)	(7.68)
PD	Population density	2.21	–0.61
		(42.57)	(21.51)
GOP	Value of per capita gross output	0.59	–0.31
		(22.48)	(18.33)
SF	Ratio of state-owned to total forestland area	0.42	–0.77
		(7.17)	(7.80)
HRS	Ratio of land under HRS to total collective land	0.79	–0.54
		(8.56)	(16.06)
U	Dummy for uncertainty	–0.09	–0.76
		(2.73)	(0.88)
HRS · *U*	Joint effect of HRS and uncertainty	–1.55	0.66
		(15.49)	(17.07)
ρ^a	Autocorrelation coefficient	0.64	0.53
	Log-likelihood	134.7	144.6

Notes: Data for managed forests are for 1957–1994 across 13 regional administrative units. Data for rainforests are for 1957–1985 across 8 administrative units. The *t*-values are in parentheses. The regional dummy variables are excluded. They are fully discussed in Zhang et al. 2000. HRS, household responsibility system.

[a]A single autocorrelation coefficient was adopted because we found that different autocorrelation coefficients did not change the estimated results or standard errors markedly.

Agricultural prices have a positive impact on forest cover in both managed and natural forests, although the effect on managed forests is statistically insignificant. This means that higher agricultural prices probably induced more land-intensive agriculture in Hainan, which means that some former agricultural land was available for conversion to forestry. In addition, higher agricultural prices would have attracted labor into agricultural production and away from timber harvest activities in both kinds of forests, perhaps because rural workers could satisfy their income needs in agricultural activities without seeking supplements from part-time timber harvesting.[3] Therefore, agricultural land use and forestland use are not in conflict. Indeed, they are complementary in some cases because forest management occurs for other reasons in addition to timber production. For example, a significant share of plantation forestry in Hainan was designed to protect agricultural lands from the downstream effects of erosion. Increasing agricultural prices encourage investment in this kind of forest.

The prices of tropical crops have a negative (but insignificant) effect on managed forests, which implies that tropical crops may compete for land with plantations. However, tropical crop prices have a positive effect on rainforest cover, and tropical crops and rainforests appear to be complements. This observation is interesting because the State Reclamation Bureau converted a large area of mature rainforest into tropical crops. Could this act have caused a decline in rubber prices and deterred the further conversion of the natural forest to tropical tree crops?

Population density has a positive effect on managed forestry and a negative effect on natural forests. Apparently, a larger population implies greater demands on the forest resource and, therefore, expansion of the managed forest combined with increasing exploitation of the natural forest. The latter may be for the products of the natural forest or for the land itself—which may be converted into other land uses. However, expanding demands for forest resources also meant more rapid development of managed forests as sources of substitute forest products and forest resource services.

General Economic Development. Economic growth had a positive effect on managed forest cover but a negative effect on the remaining natural forests. This is not surprising. With economic growth, demands for all products expand and the infrastructure improves. Improvements in roads and in the institutions that provide for and protect property rights are infrastructure changes that are particularly important for forestry. Improved roads and increasing demand support increased extraction from the natural forest and promote a decline in the natural forest cover. Improved property rights and increasing demand encourage forest management and the expansion of managed forests. In fact, the expansion of managed forests accelerated in the 1990s, just as the Hainan economy began to experience its most rapid growth. Furthermore, a wealthier society may also have greater appreciation for the nonconsumptive *in situ* benefits of forests. The fact that approximately 20% of the plantation forest in Hainan is intended for environmental purposes seems to confirm this point. The restriction on harvests from the natural forest beginning in 1985 (the last year of our regression data for that class of forest cover) seems to provide further confirmation.

Forest Policy. The share of forestland controlled by state-owned forest enterprises has a positive effect on managed forest cover, but a negative effect on natural rainforest cover. This is also expected. The state-owned forest bureaus and forest farms were set up for two purposes: to afforest wasteland and to exploit the natural forest. Of course, their logging activities in the natural rainforest were discontinued with the ban on logging the natural forest in 1985.

Decollectivization and the shift to household responsibility contracts prompted the protection and reforestation of the new household lands. Managed forest cover increased. Meanwhile, some natural forest cover was removed, at least in the early 1980s, probably to provide an alternative for the forest products that previously had come from the collective forestlands that were undergoing reforestation. That is, households apparently relied on the products of the more open natural forest while they actively improved their personal forest resources.

The uncertainty inherent in the delivery of decollectivization restrained households from reforesting as much as they might have, and the interactive term $HRS \cdot U$ shows that uncertainty had a particular offsetting effect on the improved incentives of the HRS. It caused households to increase their rate of harvest and to delay management.[4] Uncertainty in the delivery of decollectivization had little effect on the natural forest, however, because this forest remained under state management. It was not subject to decollectivization. (The coefficient associated with natural forest management is statistically insignificant.)

Accounting for the Change in Forest Cover

We can examine the relative magnitudes of impacts of the various sources of change in managed and natural forests by reviewing the elasticities and decomposing the total changes in forest cover into their sources. The coefficients for the first five variables in Table 7-1 are also their elasticities. Table 7-2 reports the decomposition analysis. (The decomposition is calculated as the estimated coefficient, which measures the effect of any change in the independent variable, times the absolute change in that variable.)

The results reported in Table 7-1 show that several factors had statistically significant effects on the expansion of managed forest cover. The decomposition analysis in Table 7-2 shows that changes in three factors—timber prices, tropical crop prices, and the institutional change characterized by *HRS*—were responsible for 64% of the increase in managed forest cover. Changes in population density and gross output explained a smaller but still important 25% of the change in managed forest cover. Population density is the only independent variable with an elastic effect (Table 7-1)—a 1% change in population density causes more than a 1% change in forest cover—but the absolute change in population density was small. Therefore, its total effect was small as well.

Timber prices and tropical tree crop prices had elastic effects on the decline in natural forest cover, and the absolute changes in these two factors were large enough over our period of analysis that they also were the largest sources of the change in natural forest cover. Population density and gross output, however, still explained 19% of the decline in natural forest cover before harvests from this class of forest were strictly limited in 1985.

Conclusions and Final Remarks

We made two key points that should be true for economic, management, and policy analyses of forestry anywhere—in China and in the rest of the world:

- managed stands and natural forests must be separated, and
- it is important to look beyond the forestry sector itself when searching for the factors that have important effects on forestry.

Table 7-2. *Accounting for the Change in Forest Cover*

	Managed forest cover			Rainforest cover		
Variable	*Estimated coefficient (1)*	*Change in explanatory variable (2)*	*Contribution to the increase (%) (3) (= 1 × 2)*	*Estimated coefficient (4)*	*Change in explanatory variable (5)*	*Contribution to the decline (%) (6) (= 4 × 5)*
p_t	0.61	26.05	15.90 (45.23)	–1.52	18.86	–28.67 (–209.82)
p_a	0.11	22.41	2.47 (7.02)	0.33	15.43	5.10 (36.28)
p_c	–0.40	28.50	–11.40 (–32.45)	1.29	10.50	13.55 (99.14)
PD	2.21	1.39	3.07 (8.74)	–0.61	1.32	–0.81 (–5.89)
GOP	0.59	9.85	5.81 (16.54)	–0.31	5.87	–1.82 (–12.96)
SF	0.42	0.18	0.08 (0.22)	–0.77	0.74	–0.57 (–4.17)
HRS	0.79	23	18.17 (51.71)	–0.54	10	–5.4 (–39.52)
U	–0.09	0	0	–0.76	1	–0.76 (–5.56)
HRS · U	–1.55	0	0	0.66	10	6.60 (48.31)
Residual			1.05 (3.00)			–1.25 (–9.14)
Total			35.14 (100)			–14.04 (100)

Note: See Table 7-1 for a description of the variables. Because the regions and time covered in the two regressions are different, the change in explanatory variables is not same between the managed forest and rainforests.

Then we illustrated these points with evidence from the island province of Hainan.

It is important to separate managed forests and forest plantations from unmanaged natural forests when examining market and policy effects on forests and also when considering the resource and environmental products and services the forests provide. Markets and policies can have opposite effects on these two classes of forest, encouraging expansion in one and contraction of the other. Both classes of forest can be sources of industrial timber and, depending on their location, both can enhance watershed values, but natural forests have distinct advantages for the protection of many other environmental services. The distinctions between the two classes of forest also imply that any potential for substitution between the two takes on special meaning. In particular, larger managed forests mean greater opportunities to satisfy demands for timber without interfering with the natural forest and its special nonconsumptive environmental services.

When considering market and policy influences on the forest, it is important to search broadly. Direct forest policies can be important, but general economic conditions such as macroeconomic growth and stability may also be important, especially in markets with forestry sectors that are small relative to all other economic activity.

In Hainan, seven of the nine explanatory variables—including the critical timber price variable—had opposite effects on the levels of managed and natural forest cover. The prices of timber and of tropical forest products and the HRS institutional change that was closely associated with forestry itself had the largest absolute effects on changes in the two classes of forest cover. However, agricultural prices and the level of general economic activity were also important. Agricultural prices explained 7% and 35% of the changes in managed and natural forest cover, respectively. The level of general economic activity explained 17% and 13% of the changes in managed and natural forest cover, respectively.

Two decades ago, the area of unforested but potentially forestable land in Hainan was as large as the total forested area. Since then, general economic growth, the development of the export market for wood chips, and favorable institutional change have all had favorable impacts, and managed forest has gradually replaced the potentially forestable area. Some of the new managed forest has been for commercial timber, and some has been reforested primarily to protect local environmental values.

Hainan's experience is illustrative of what we might anticipate for the rest of China as well. China today is very much like Hainan two decades ago. Its significant share of degraded forestland is not due to agricultural occupation or to other land uses but to the lower timber prices that are the result of weak demand from a less-developed economy, poorly defined land tenure, and insecurity in the existing property rights. Most of the potentially forested land does not lack a defined owner. Rather, the owner is unable or unwilling to enforce the property rights to the land. This situation makes the lands effectively open-access resources on which inactive management is a rational choice.

Therefore, the biggest challenge to China's forestry is not population growth and economic growth, and it is not poor land quality or limited land availability.

It is weak demand and the weak institutions that make forest management costly. The institutions that are important for reducing transactions costs may be even more important than price incentives for the transition from harvesting natural forests to active forest management. Therefore, future policy should emphasize the development of institutions that can reduce the protection costs for property rights to forest management and provide greater security and confidence for farmers regarding the continuity of their property rights.

The current massive reforestation and afforestation programs of the central government and some international donors still focus on direct public investment. This may not be the most effective approach. Closed access is infeasible without the institutions that protect exclusion inexpensively. Without closed access, any investment in silviculture is likely to be wasted. If forest management is profitable, private investment will follow. The priority of government should be to invest in the institutions that promote the definition, transference, and protection of property rights.

Notes

1. We also attempted to estimate the sources of change in the levels of natural and managed forest cover. However, the natural forest regression was unsatisfactory and, without it, any comparison of effects on the two classes of forest is impossible. The natural forest regression was unsatisfactory because the very small intertemporal changes in the dependent variable created an insurmountable multicollinearity problem as well as coefficients that were statistically insignificant.
2. Lagged values of the exogenous variables did not improve the statistical performance of the estimated equations.
3. Bluffstone (1995), Foster et al. (1997), and Hyde and Kohlin (2000) discuss the theory and provide similar examples from India and Nepal.
4. This is further confirmation of the critical impact of uncertainty on forestry. Liu and Edmunds introduced the idea in Chapter 2, and Yin demonstrated it with empirical evidence in Chapter 4.

References

Bao, K. 1991. The Provincial Government's Making an Initiation: Experiment of Sustainable Development in Hainan. In *Proceedings of International Advisory Meeting on the Economic Development of Hainan in Harmony with the Natural Environment*. Haikou, China: International Advisory Council for Hainan Economic Development, 93–96.

Besley, T. 1995. Property Rights and Investment Incentives: Theory and Evidence from Ghana. *Journal of Political Economy* 103(5): 903–937.

Bluffstone, R. 1995. The Effect of Labor Markets on Deforestation in Developing Countries under Open Access: An Example from Rural Nepal. *Journal of Environmental Economics and Management* 29(1): 42–63.

Chen, Z. 1948. *The Exploitation of Hainan's Resources* [in Chinese]. Haikou, China: Zhongzhen Publishing House.

Foster, A., M. Rosenzweig, and J. Behrman. 1997. Population and Deforestation: Management of Village Common Land in India. Draft manuscript. Philadelphia,: University of Pennsylvania, Department of Economics.

Greene, W. H. 1997. *Econometric Analysis,* 3rd ed. New York: Prentice-Hall International, Inc.

Hainan Forestry Bureau. 1957 through 1995. *Hainan Forestry Statistical Materials* [in Chinese]. Haikou, China: Hainan Forestry Bureau.

Hainan Statistics Bureau. 1957 through 1995. *Hainan Statistical Yearbook* [in Chinese]. Haikou, China: China Statistical Press.

HAZC (Hainan Agricultural Zoning Commission). 1980. *Collection of Hainan Agricultural Zoning Reports* [in Chinese]. Haikou, China: HAZC.

Hyde, W., and G. Kohlin. 2000. Social Forestry Reconsidered. *Silva Fennica* 34(3): 285–314.

Hyde, W., G.S. Amacher, and W. Magrath. 1996. Deforestation, Scarce Forest Resources, and Forest Land Use: Theory, Empirical Evidence, and Policy Implication. *World Bank Research Observer* 11: 223–248.

Kmenta, J. 1986. *Elements of Econometrics*, 2nd ed. New York: Macmillan.

Mather, A. 1990. *Global Forest Resources.* Portland, OR: Timber Press.

Rudel, T. K. 1998. Is There a Forest Transition? Deforestation, Reforestation, and Development. *Rural Sociology* 63(4): 533–552.

Situ, S. 1992. *Historical Land Use Development in Hainan* [in Chinese]. Haikou, China: Hainan Press.

Walker, K.R. 1965. *Planning in Chinese Agriculture: Socialization and the Private Sector 1956–1962.* London: Frank Cass and Co.

Yin, Runsheng, and David Newman. 1997. Impacts of Rural Reforms: The Case of the Chinese Forest Sector. *Environment and Development Economics* 2(3): 291–305.

Zeng, Q. 1994. A Brief Introduction to Closing the Land for Reforestation in Hainan Province. *Tropical Forestry Newsletter* 2: 24.

Zhang, Y. 2000. Deforestation and Forest Transition: Theory and Evidence in China. In *World Forests from Deforestation to Transition?* edited by Matti Palo and Heidi Vanhanen. Dordrecht, the Netherlands: Kluwer Academic Publishers, 19–43.

Zhang, Y., G. Dai, H. Hung, F. Kong, Z. Tian, X. Wang, and L. Zhang. 1999. The Forest Sector in China: Towards a Market Economy. In *World Forests, Society, and Environment,* edited by Matti Palo and Jussi Uusivuori. Dordrecht, the Netherlands: Kluwer Academic Publishers, 371–393.

Zhang, Y., J. Uusivuori, and J. Kuuluvainen. 2000. Econometric Analysis of the Causes of Forestland Use/Cover Change in Hainan, China. *Canadian Journal of Forest Research* 30: 1913–1921.

CHAPTER 8

Forestry, Poverty, and Rural Development

Perspectives from the Bamboo Subsector

Manuel Ruiz Pérez, Brian Belcher, Maoyi Fu, and Xiaosheng Yang

THIS CHAPTER AND THE NEXT examine two issues, income distribution and the environment, that are increasingly important in contemporary China as a result of 20 years of successful reforms and economic growth. Rural incomes have increased more than 600% since 1978, but more rapid growth in China's urban areas and especially around its coastal cities has created an expanding disparity between the incomes of urban and rural households and particularly between western China and the rest of the country. China's leaders demonstrated their concern by introducing the Western Regional Development Program in 1997 to address the imbalance.

Meanwhile, China also experiences several environmental problems related to its rapid growth: deteriorating air quality in the cities, deteriorating water quality downstream from many industrial sources, and continued soil erosion, an age-old problem. China's leaders are addressing these environmental problems as well. Premier Zhu Rongji, in the Government Report to the People's Congress on March 5, 1999, identified sustainable development as one of two fundamental government policy strategies for the twenty-first century. President Jiang Zemin stressed the importance of environmental protection at the annual workshop on Population, Resource, and Environment on March 13, 1999, when he announced that any enterprise not in environmental compliance by 2000 would be closed.

We are most interested in forestry's role with respect to these issues, particularly its potential for helping overcome poverty and income disparity and its function as a source of both environmental loss and environmental recovery. We will examine the first issue in this chapter, and Sayer and Sun will examine a component of the second in Chapter 9. Both topics include an extensive range of issues, yet little empirical research has been conducted on either. Therefore, we will rely on evidence from our own research in the bamboo subsector of forestry to illustrate the impacts of forest development on rural households and, in Chapter 9, Sayer and Sun will highlight their concern for the loss of natural forest habitat as an example of the very serious environmental consequences of China's forest transition.

This chapter begins with a summary of the state of rural poverty in China and the potential role for forestry. It continues with a general discussion of the bamboo sector and then presents evidence from our own research regarding the household income effects of the rapid expansion in bamboo production. Bamboo in all of its many uses is the third most important forest product in China (after fuelwood and commercial timber). It is a particularly useful product for our inquiry because it has not been subject to as many policy restrictions as commercial timber. Therefore, its recent history is more representative of the impacts that unfettered forestry-sector development could have on household incomes and general regional development. We will observe that many households have benefited from the expansion in bamboo production, but the poorest have not been the greatest benefactors. Furthermore, bamboo seems to have a role in development, but its importance declines as development progresses. These observations on bamboo do not reject the standard arguments about forestry as a development leader. However, they do prompt us to urge caution in their applications, particularly with regard to assistance for the very poorest.

Poverty and Forestry

Despite China's rapid growth, tens of millions of its people are still poor. In 1998, 230 million people, 18.5% of the population, lived below the US$1 per day international standard of poverty. Even under China's narrower standard of income poverty, 60 million people, 4.6% of its population, are considered absolutely poor (World Bank 2000).[1]

Poverty is more pronounced in the rural areas. Average rural per capita income is less than 40% of the urban average, and rural living expenditures are less than 35% of urban expenditures per capita (*China Statistical Yearbook* 2000). The nature of poverty is different in rural and urban settings as well. Because the land distribution component of China's reforms was conducted in an egalitarian manner within each village, rural poverty has less to do with food security than with a generally lower level of social services and a higher level of vulnerability to natural hazards and macroeconomic fluctuations (Chapters 1 and 2, Jalan and Ravallion 1999, Liu et al. 1999).

Economic development has been a source of increasing household and regional differences. For example, per capita gross domestic product (GDP) in southwestern China is only 44% of that in the eastern coastal areas, and that gap has been increasing (*China Statistical Yearbook* 2000). For another example, minority nationalities account for only 9% of China's total population, but they account for around 40% of the population in absolute poverty (World Bank 2000).

Remoteness and isolation are correlated with poverty. In the mid-1990s, 496 of the 592 officially designated poverty-stricken counties were in mountainous regions (MOF 1995). It is frequently in these less accessible areas, where natural constraints limit agriculture, that forests have been preserved. Therefore, the pattern of isolation, poverty, and extensive forest cover is common, and disentangling the causal links is a difficult task.

The poor performance of the forest industry relative to China's full industrial sector compounds the situation (Research Group of Forestry Economics 1998). The Third National Industry Survey showed that financial losses are common in all forest enterprises, but state-owned forest enterprises (SOFEs) fare the worst. The average deficit in the SOFEs was 13.9 percentage points higher than the average for all industrial sectors (SFA 1999). Forest workers have among the lowest standards of living in China, and their salaries appear at the bottom of the list of more than 60 sectors published in the national statistics. The average forest worker receives only 55% of the national average salary (Chen and Lu 1999, *China Statistical Yearbook* 2000).

At the same time, forestry often represents the main—and sometimes the only—income-generating activity in many poor regions. At a national level, forestry contributes only 2% to farm-derived rural cash income (*China Statistical Yearbook* 2000). However, in some of the income-poor but forest-rich counties, forestry is the source of up to 80% of household revenues (Da 1999, Zhang 2000). Indeed, forest industries are often the only industry in many poor counties. In these cases, they are both a pillar for the development of the local economy and a critical taxpayer (Peng 1999, Zhang and Yuan 1999).

Some argue, with reference to the high correlation between the occurrence of forests and poverty, that the depletion of forest reserves aggravates poverty (Niu and Harris 1996, Smil 1997, Harkness 1998). Others see it differently. Li and Veeck (1999) explored the contribution of forests to rural income in poor areas with multicounty economic models based on county and provincial statistics. They found significant overlap between counties officially classified as having abundant forest resources (*senlin fengfu xian*) and those classified with severe poverty (*pinkun xian*), yet no relationship between forest variables and income. They concluded that the availability of forest resources contributes only slightly to per capita income.

Therefore, opinion is divided, and the question remains: do forest resources and forestry activities promote rural development, or are they part of the poverty trap? We will explore this issue with reference to the bamboo subsector.

General Forestry Opportunity and the Bamboo Subsector

Why is bamboo an appropriate focus for our inquiry? Bamboo has comparative advantages over commercial timber: a short production cycle, expanding demand, relatively unregulated harvest and shipment, and low taxation. It is an alternative investment for farmers who recognize the potential gain from commercial timber investments but find timber operations highly regulated. The regulations reduce timber's financial returns and also make them less certain. Farmers are more confident that they can claim the eventual gains from investments in bamboo for themselves. In Chapter 2, Liu and Edmunds observed farmers investing in bamboo for this reason in Hunan and Zhejiang. In Chapter 3, Liu et al. observed the same behavior throughout the Southern Collective Forest Region. Others have observed it as well (Joint Survey Group 1999, Wang and Zhao 1999, Cao and Zhang 1997). It suggests that our observations of investment behavior and income returns in bamboo are indicative of what we might expect if China loosened its regulations on harvest levels and shipments for commercial timber.

Private Forest Opportunity

China has one of the lowest per capita forest endowments in the world—about one-half the world's average. This implies that the accessible natural forest is small relative to the demand for forest products and, therefore, it is not surprising that China relies on managed stands for a large share of its forest production. Farm production of forest products is unusually important.

This point affects forest investment and development planning, and it has a direct effect on how forestry contributes to household incomes and poverty alleviation. Forestry departments at different administrative levels, as sole authorities or in conjunction with other departments, have the mandate to implement development plans on forestland. However, their regulations constrain timber harvest levels and shipment, even from farmlands, and the taxes and charges they impose are further disincentives. Furthermore, forestry departments generally permit the conversion of designated forestland to other uses only with the (unlikely) permission of the local planning authority. Therefore, farmers looking for opportunities to improve the returns on their lands within the restrictions of the law often turn to crops such as bamboo and fruit trees.[2] China classifies these crops as "economic forestry" in its unusually broad definition of forests.

In sum, this means that farmers, forest farms, and the forest authorities actively engage in the expansion of orchards and bamboo plantations on forestland, auctioned wasteland, and even some former cropland (Hanstad and Li 1997, Yu et al. 1999). Economic forestry expanded from 6.1 million ha with 7 million metric tons of output in 1978 to 21.9 million ha with 53 million metric tons of output in 1997. China ranks first in the world in the area and output for these crops. Similarly, bamboo plantations increased from 3.2 million ha and 4.4 million metric tons of output in 1980 to 4.3 million ha and 14.2 million metric tons of output in 1999 (Ruiz Pérez et al. 2001). Thus, while conventional timber forests have been depleted and the timber-based forest industry shows signs of stagnation, economic forests in general and bamboo plantations and their associated industries in particular are thriving, offering farmers good opportunities to increase their incomes and to emerge from poverty.

The Bamboo Subsector

The remainder of this chapter traces the recent development of China's bamboo sector beginning with a discussion of bamboo itself and continuing with our research experience, first in Anji County and then in an extended comparison of six counties.

Between 300 and 500 species of bamboo (depending on the taxonomic criteria) grow in China. More than 100 have commercial value (Zhu et al. 1994). In the mid-1990s, China had 7 million ha of commercial bamboo forest, 4 million ha in plantations, and the remainder in natural stands. The most common species, moso bamboo (*Phyllostachys heterocycla*), occupied 2.8 million ha. By 1999, bamboo accounted for 3% of China's total forest cover, but contributed about 25% of the value of China's forest exports. The total value of unprocessed and processed bamboo products was 23.14 billion yuan (US$2.8 billion) (SFA 1999).

Bamboo's uses fall into two main categories. The culms (stems) are used in timberlike applications in unprocessed (poles and scaffolding) or processed (panels, bamboo flooring, pulp, and paper) forms. Edible bamboo shoots are used as a vegetable, fresh or in various processed forms.[3] Because of this dual use, bamboo has economic characteristics of both conventional timber-oriented forests and fruit-bearing trees.

Bamboo plantations have several attractive economic attributes. Their investment costs are low, their production cycle is annual or biennial, and their multiple products allow for flexible market responses. Culm production has a longer production cycle than shoot production. It provides a generally stable core economic activity with reliable financial returns. For moso bamboo, shoots are a biennial crop. The financial returns are often greater than those of culms, but the prices of moso shoots are also more volatile, and it is easier to expand or contract shoot production in response to changing market conditions. Bamboo plantations are more profitable than industrial timber, generating between two and five times the income per unit of land generated by Chinese fir or pine plantations. Consequently, when environmental conditions allow, farmers find bamboo more attractive than timber plantations and they actively replace conifers with bamboo.

The logging ban imposed in 1998 increased the demand for timber substitutes, and bamboo production benefited immediately. Bamboo prices were still increasing at a 5–10% annual rate in 1999, and new bamboo plantations increased by almost 17% in the same year. Culm production has increased by 30% since 1997. Some expect that bamboo will substitute for 29 million m^3 of wood by 2010 (CFIC 1998), and some at the State Forestry Administration believe that bamboo could become the dominant forest industry in China by 2050 (Li and Xu 1998).

Anji's Experience[4]

Anji County is in Zhejiang, an economically dynamic coastal province near Shanghai. It is one of four provinces that account for two-thirds of China's bamboo plantation area, and one of ten provinces known as the "native place of bamboo." The province in general and its bamboo sector in particular began developing earlier and more rapidly than some others, but its pattern of development is a pattern that other bamboo producing provinces have followed since, and it is a pattern that we might anticipate for forest development in general.

In 1975, Anji had 51,400 ha of bamboo, almost 99% of it managed by local collectives with the small remainder managed by state farms. Bamboo was sold through the state marketing cooperative (a monopoly) at a fixed procurement price. By 1998, primary bamboo production accounted for 70% of the forest output value in Anji and 15% of total farm income. The bamboo-processing sector accounted for 67% of the county's total industrial output, a very substantial increase from only 8% in 1980.

Beginning in 1980, the government allowed market sales of all output in excess of production quotas. The productive area in bamboo immediately increased almost 5%. However, the sharpest increase in Anji's production came in 1983–1984, with the adoption of the household responsibility system (HRS). The collectives managed 43,000 ha of moso bamboo in 1982 but only 3,400 ha by 1984. Individual farm household management increased from zero in 1982 to 40,000 ha by 1984. Total output in the county increased 63% between 1983 and 1988, mostly due to intensification and with little expansion in the land area devoted to bamboo. Average production increased from 208 culms/ha in 1975 to 224 culms/ha in 1982 to 333 culms/ha in 1989.

Market prices were higher than procurement prices during the early reform years. They were an incentive for farmers to bypass the quotas and the lower procurement prices, and most of the increase in production after the introduction of HRS was sold at the higher market price. This forced the government to continually revise its procurement prices upward until the two-price system was eliminated in 1985. By this time, individual farmers were responsible for 91% of the county's total production of bamboo.

Prices and production continued to increase until 1990, when excess supply caused prices to decline temporarily. Production declined 12% in response but subsequently more than recovered, increasing another 29% by 1994. Altogether, prices increased almost 300% (in constant terms) between the late 1970s and mid-1990s. The production of culms increased 79% over the same period and 90% by 1998. Total production increased even more if the output of bamboo shoots is included.

The Development of the Processing Sector. In 1978, 96% of Anji's bamboo was sold in unprocessed form and shipped out of the county. Nineteen local bamboo-processing establishments employing 460 workers produced three general categories of processed goods—mats and flooring, shoots, and handicrafts and furniture—worth only 960,000 yuan (US$670,000).

Local processing capacity grew rapidly, however. The initial reforms in the marketing system that allowed market sales of production in excess of quotas provided a basis for many new nonstate (township and village enterprise or TVE) processing facilities. The TVEs also provided an outlet for China's vast accumulation of private savings. By 1985, Anji had 154 bamboo-processing establishments that employed 3,370 workers and produced 12.31 million yuan (US$4.16 million) of annual output.

Joint ventures with foreign investors were approved for operation in Anji in 1988, and Chinese participants were permitted to retain their foreign currency earnings. This reform fueled a second round of expansion in processing capacity. Nineteen new joint ventures employed more than 3,000 workers and produced 36 million yuan (US$4.5 million) of annual output by the mid-1990s. Much of their output was exported, as exports from domestic operations and joint ventures combined exploded from US$100,000 in 1980 to $23.4 million in 1994 to $56.3 million in 1998.

The joint ventures introduced new physical capital and improved technologies, creating a wider range of products and enhancing product quality. Competitiveness and opportunities in external markets improved, and they created a local multiplier effect that increased the demand for culms and shoots and encouraged the establishment of many small household preprocessing units in support of some of the larger joint ventures.

Rapid growth in the processing sector put upward pressure on the prices of the primary bamboo resource until the mid-1990s. In response, the marketing system gradually grew more specialized. Marketing intermediaries began appearing until more than 200 bamboo traders were operating by the mid-1990s. Meanwhile, some processing establishments, concerned for the supply of their own raw materials, began arranging production agreements with farmers at prices agreed upon before the harvest period. Some agreements even included cash advances. Eventually, the processing sector began importing raw material from other counties.

By 1998, 1,182 processing establishments employing 18,914 workers were operating in Anji. Annual production exceeded 875 million yuan (US$105 million) in value. However, the entire increase in processing between 1975 and 1998 occurred while the number of bamboo farmers expanded only moderately from 111,000 to 123,000. By 1998, infrastructure development and specialization in processing had developed to the point where Anji's farmers shipped one-third of their unprocessed bamboo (mainly low-value small culms and branches) out of the county, while the processing industry imported an amount equal to 20% of local raw material production.

The sector went through a period of stagnation in the mid-1990s, but it has since recovered. First, the new supplies of raw materials from outside the county dampened the upward pressure on raw material prices, then the East Asian financial crisis of 1997 decreased export demand. The processing sector experienced a period of overcapacity and adjustment. It recovered in the late 1990s, and growth has continued, although at a diminished rate. The sector we observe today, including both its raw material and its processing components, could probably be called "mature."

Bamboo and the Distribution of Farm Income.[5] The growth in primary bamboo production over the period of China's reforms has been impressive, but it has been achieved with little expansion in the number of farm households or area managed. The sector has diversified from a strictly farming activity to incorporate a full range of marketing and processing activities that employed almost 19,000 new workers in the county by 1998. However, the bulk of those affected by the rapid expansion of Anji's bamboo sector remain the more than 120,000

Table 8-1. *Basic Township Data, 1994–1995*

Township	*Population*	*Total area (mu)*	*Population density per km²*	*Riceland (mu)*	*Rice (mu/person)*	*Bamboo (mu)*	*Bamboo (mu/person)*
Kuntong	16,155	135,036	179.5	5,571	0.34	68,633	4.25
Fenghuang	14,918	130,510	171.5	5,928	0.40	57,496	3.85
Gangkou	9,783	71,609	204.9	4,737	0.48	48,067	4.91
Tianhuang	19,001	151,952	187.6	11,475	0.60	64,848	3.41
Baofu	18,316	200,636	136.9	8,303	0.45	94,506	5.16
Zhangchun	15,656	131,071	179.2	6,867	0.44	64,728	4.13
Yonghe	13,463	162,980	123.9	5,337	0.40	80,225	5.96
Chiwu	12,479	111,550	167.8	6,092	0.49	47,896	3.84

Note: Mu is a traditional measure of land area that varies according to land use. For the normal mu used here, 15 mu = 1 ha.
Source: Anji Bureau of Statistics.

bamboo farmers themselves. We will examine the effects of the sector's growth on their incomes.

Our analysis is based on several years of research. We began with discussions with key informants across the county, and followed these discussions with a household questionnaire inquiring about land status and land use, income sources, labor, and expenditures. Two hundred survey participants were selected from a stratified random sample of households from five villages per town in each of the eight towns in the county. We assembled the data from the questionnaires in biennial increments corresponding with bamboo's two-year production cycle, and we supplemented the survey data with evidence from farm records since 1989–1990. We feel confident of farm records dating forward from that time.

Tables 8-1 and 8-2 summarize the basic township data and characterize the typical farm household in 1994–1995. Average farm household income was 14,033 yuan. Farm income derived from four principal sources: rice production, livestock, bamboo production, and off-farm employment. Bamboo was the second most important source (after off-farm employment). It provided 25% of average household income, 21% from culms, 2.5% from shoots, and 1.5% from branches. The 1989–1990 farm records do not provide the full data for comparison, but we know that bamboo provided 24% of farm household income in 1989–1990, and household incomes in the mountainous regions of Anji rose 37% in real terms between 1989–1990 and 1994–1995. The rate of income increase among our households (6.7% per year) was slower than for all Chinese households but more rapid than for all farm households.

We separated the 200 farm households into five income classes to examine whether the importance of bamboo changed with the level of farm income. Figures 8-1 and 8-2 describe the results for each income quintile for 1989–1990 and 1994–1995, respectively. Both figures demonstrate a convex shape, which means

Table 8-2. *Basic Data of the Farmer Sample in Anji County, 1994–1995*

Variable	*Mean*	*Standard deviation*	*Minimum*	*Maximum*
Age, head of household (years)	43.6	8.88	24	69
Family size	4.0	1.05	1	7
Workers per family	2.6	0.89	1	6
Education level [a]	2.1	0.74	1	3
Arable land (mu)	2.7	1.83	0.2	18
Cash trees (mu)	0.4	1.53	0	14
Bamboo (mu)	14.9	10.05	1	70
Other forest (mu)	3.5	10.49	0	111
Total land (mu)	21.2	15.64	2.8	127

[a]Education level in three categories: 1 = high school (12 years); 2 = secondary school (9 years); 3 = primary school (4 years).

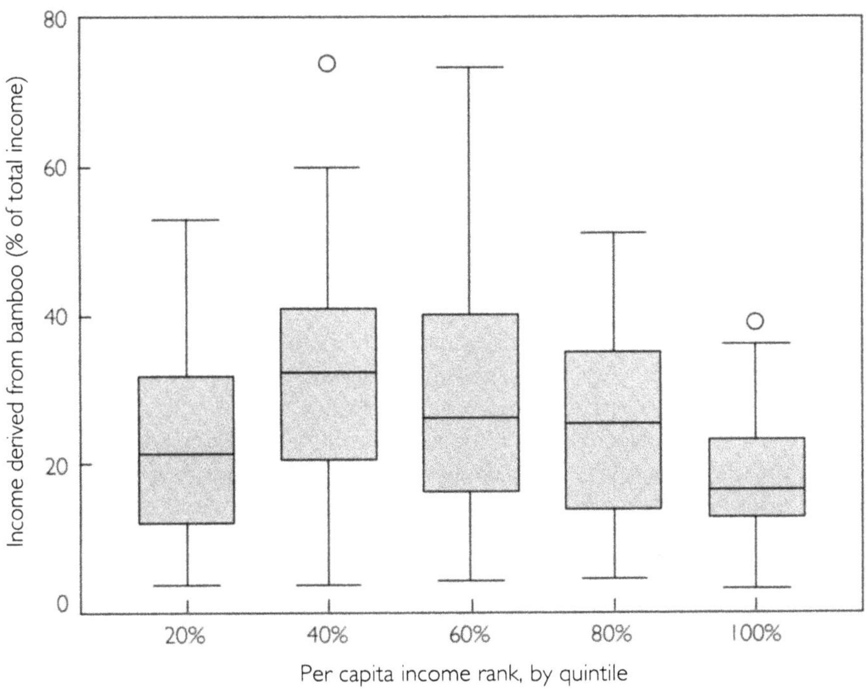

Figure 8-1. *Relative Importance of Bamboo Income in Anji County, by Household Income Class, 1989–1990*

Note: Median, interquintile range, and extreme values for the contribution of bamboo income to total income in each quintile.

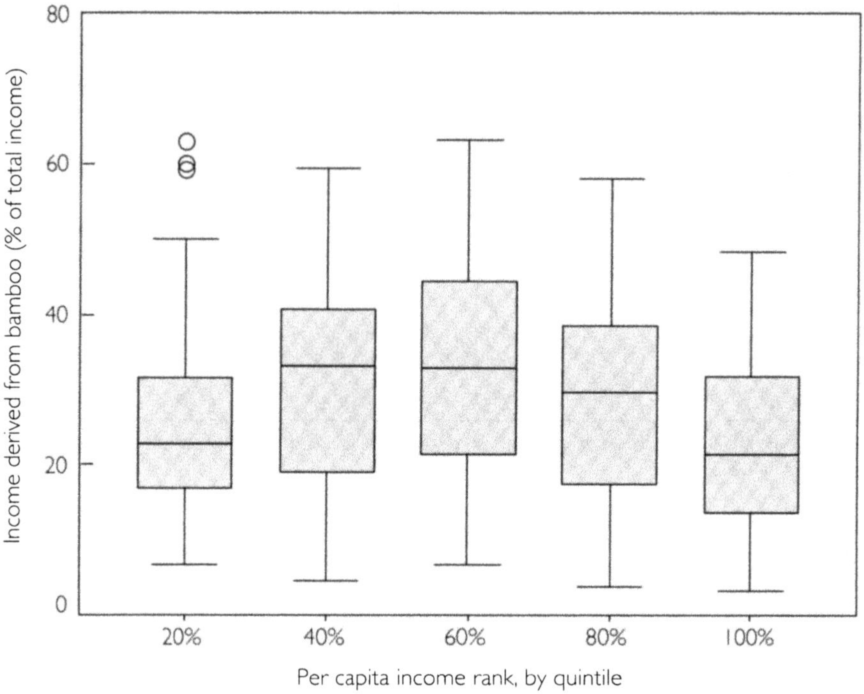

Figure 8-2. *Relative Importance of Bamboo Income in Anji County, by Household Income Class, 1994–1995*

Note: Median, interquintile range, and extreme values for the contribution of bamboo income to total income in each quintile.

that bamboo was a more important source of income for middle-income quintiles—providing up to 30% of household income for the second quintile in 1989–1990 and the third quintile in 1994–1995. Bamboo was the source of barely 20% of total income for the extreme quintiles in either period.

The 1994–1995 survey shows that agricultural income was relatively constant across all quintiles. Therefore, we conclude that agriculture provided a common economic foundation for all farm households. Income from bamboo increased in absolute terms (but not relative terms) for higher quintile households. It was an increasingly important source of on-farm income for higher quintile households. However, off-farm employment was the largest source of the income difference between households in the different quintiles, and it was especially important for households in the highest income quintile.

We are at least as interested in what explained the growth in household income. This is the information that will show whether bamboo helped increase household income and helped the poorest households escape poverty. We examined the sources of income growth by regressing the difference in per capita income over the five-year period on various measures of household labor and land resources. Table 8-3 reports the results. Clearly, male labor, arable land, bam-

Table 8-3. *Sources of Difference in Per Capita Income in Anji County, 1989–1990 to 1994–1995*

Variable	*B-coefficient*	*β-coefficient*	*T-value*
Age, head of household (years)	–19.8	–0.127	–1.996
Family size	–297.8	–0.225	–3.198
Male/total labor ratio	1,361.6	0.152	2.244
Arable land	114.6	0.151	2.373
Bamboo land	53.9	0.390	5.997
Off-farm/total work ratio	855.2	0.157	2.420
Township	107.5	0.178	2.717
Constant	1,632.8		2.863
$R^2 = 0.341$	Adjusted $R^2 = 0.317$	$F = 14.178$	Probability $F < 0.0001$

boo land, and off-farm employment had critical income-increasing effects. The income elasticities of bamboo land and off-farm labor are 0.32 and 0.48, respectively. They mean that a 1% increase in bamboo land increased household income only 0.32% for the average family, while a 1% increase in the ratio of off-farm employment to farm employment increased household income 0.48%. Once more, it is clear that the improvements in bamboo production opportunities benefited household incomes, but additional off-farm labor opportunity was the crucial ingredient for the significant income growth that occurred over the period between 1989–1990 and 1995–1996.

Multicounty Comparison of Bamboo Development

Anji is an interesting case because it illustrates a complete pattern of development from the expansion of production of the primary resource to the early development of a resource-processing sector to the sector's full diversification, technological advance, and participation in global markets. The period of this pattern corresponds to the 20-odd year period since the beginning of China's economic reforms. However, Anji is only one county, and confirmation (or rejection) of its growth pattern with evidence from other regions would provide a better basis for conclusions that can be generalized about bamboo or broader forest production, development, and impacts on household incomes.

Therefore, we extended our earlier research to include six counties in three provinces along the east–west gradient of China's subtropical bamboo producing region (Anji and Longyou in Zhejiang Province, Pingjiang and Taojiang in Hunan Province, and Muchuan and Changning in Sichuan Province [Figure 8-3]). The eastern counties are closer to the larger domestic markets and also to the coast and its export opportunities. Household incomes are higher in the east, and the two eastern counties have grown more rapidly than the other four. In fact,

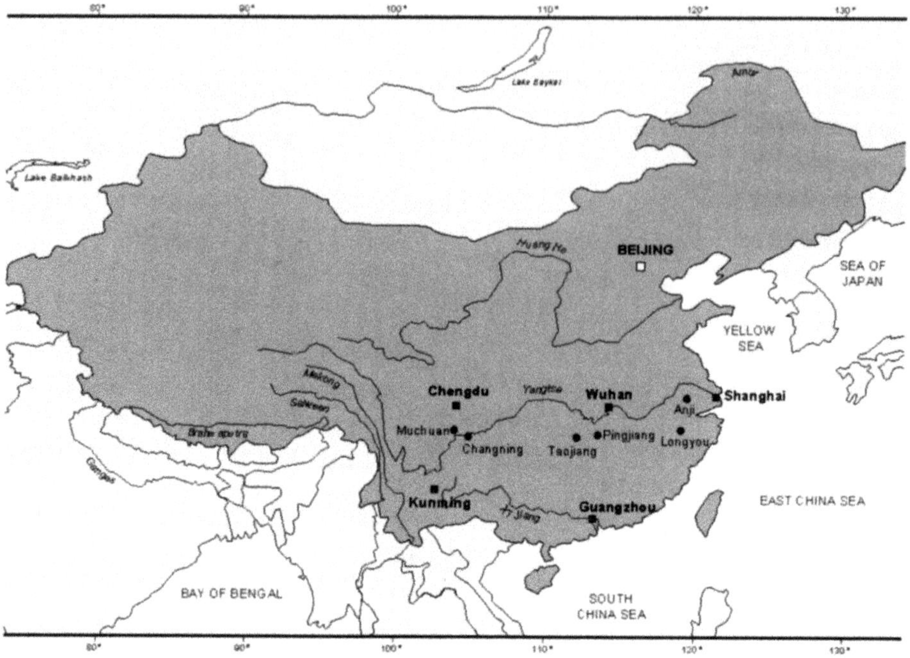

Figure 8-3. *Areas of Bamboo Research*

Note: Circles indicate counties where research was conducted; dark squares indicate provincial cities.

GDP growth in these two counties was even more rapid than national growth between 1980 and 1998. This eastern growth has contributed to the increasing regional income differences that have become a major new policy concern in China and one justification for this chapter.

On the other hand, China has actively invested in infrastructure over the past 15 years. This investment, especially in highways and railroads, and continued economic reform should have improved the competitive position of producers in the less accessible western regions. Where development in western China has followed Anji's experience, only with a lag in time, we have the information to draw generalizable conclusions about bamboo, development, and impacts on household incomes. Otherwise, observations different from or even contrary to those of Anji would raise new questions and prompt us to use caution in making generalizations from Anji experience. Observations on the pattern of development have become more critical since China embarked on its aggressive Western Regional Development Program in 1997. Evidence that conforms with the pattern of development observed in Anji will allow us to anticipate some of the potential future effects of this new program.

Our assessment of the six new counties is based on county-level statistics from 1980 to 1998, interviews with key informants in all six counties, and detailed surveys of farm households and industrial firm managers in three of those counties (Longyou, Pingjiang, and Muchuan).[6]

Primary Production[7]

In general, bamboo prices increased relative to the national price index for all commodities from the mid-1980s until 1994. They provided a signal for expansion and intensification (Gu 1992). Price stagnation in 1994 deterred further growth in the sector until the logging ban in 1998 provided a new stimulus for bamboo as a substitute for industrial timber. The number of farmers growing bamboo in the six counties increased from 244,000 in 1980 to 465,000 in 1998 as the share of all farmers growing bamboo expanded from 19% to 26%.

Production increased in each of the six counties, and key informants in all six identified the stable policy environment as a precondition for expansion. (This observation supports a central point of Chapters 2 and 4.) Culms were the dominant product in all three regions (from moso bamboo in the east and center, and from *Sinocalamus affinis* in the west), but all counties have begun diversifying and expanding their production of bamboo shoots. Land opportunity costs were greatest in the two richer eastern counties. Both the land area in bamboo and the number of farmers growing bamboo expanded least in these counties. The eastern counties are mountainous, and the expansion that occurred in these counties was assisted by regulations restricting agriculture and encouraging forestry on slopes greater than 25%. The central county of Taojiang suffered a bamboo locust plague in 1983 that affected 25% of the county, sharply reducing its production and limiting the expansion of its land area in bamboo for several years. The land area in bamboo expanded most rapidly in the two western counties.

Productivity per land unit was higher in the eastern counties than in the central counties, and the expansion in eastern bamboo production was almost entirely due to intensification. The richest county, Anji, experienced the highest average productivity (25 culms/mu or 315 culms/ha), and the poorest, Pingjiang, experienced the lowest (only 7 culms/mu or 105 culms/ha). This difference is consistent with the expectation of greater farm specialization, capital availability, and opportunity costs of land and labor in the richest counties, where farmers invest more in their bamboo plots.

Bamboo production expanded most rapidly in the two western counties, although productivity is not strictly comparable between the west and the other two regions. The species typically planted in the west cost less to establish and have a shorter production cycle than that planted in the east.

The Processing Sector

The processing industry expanded in five of the six counties, offering opportunities for farmers to increase their bamboo production and obtain additional income from household-based preprocessing as well as off-farm employment in the processing industry (Table 8-4). However, the profile of the processing industry differed markedly across counties, both in its 20-year growth trend and its current composition. We will review the experience of those three counties (Muchuan in the west, Taojiang in the center, and Longyou in the east) for which we have the benefit of survey responses from enterprise managers.

Table 8-4. *Percentage of Local Processing of Bamboo Culms*

	Local processing (%)		
County	*1980–1985*	*1995–1998*	*Increase*
Anji	14	62	48
Longyou	10[a]	71	61
Pingjiang	41	23	−18
Taojiang	38	100	61
Changning	23	88	65
Muchuan	100	100	0

[a]Data for 1983–1985 only.

Muchan. Muchuan's industrial sector is not diverse, neither generally nor in its bamboo processing. Paper accounted for 99% of all processed bamboo products and 36% of all county industrial output in 1998. The many small-scale, labor-intensive, private mills that produce specialty ceremonial papers characterize the industry, and they have been responsible for the sector's greatest growth over the past 20 years. Nevertheless, the county does contain a few mechanized mills, one of which increased its capacity to 60,000 tons/year in 2001.

The high rate of profit in Muchuan's paper industry (24%) indicates an industry that is growing and that will continue to expand. In fact, we should expect continued growth because China's aggregate growth has been so strong and sustained and because the paper industry generally grows faster than per capita income growth in most countries.

Taojiang. Taojiang's economy is somewhat more diverse than that of Muchuan or Longyou. Bamboo processing made up only 12% of total industrial output in 1998. Bamboo mats, an industry that did not exist in Taojiang before 1992, comprise 84% of all processed bamboo output. The industry's 19% profit rate indicates that it is still growing and probably will continue to grow.

Of the three counties, Taojiang has the lowest wage, lowest per capita income, and highest contribution of the primary sector to county GDP. The opportunity for the processing sector has been good. The industry can take advantage of low wages, and it has expanded more rapidly than the industry in Muchuan or Longyou.

Longyou. Longyou is the most developed of the three counties. Its per capita income is approximately twice that of Muchuan or Taojiang, and its industrial structure is more diverse. Bamboo processing comprised only 7% of total industrial production in 1991, and it has declined ever since. The paper industry dominates the bamboo sector with 54% of total output, but the plybamboo and bamboo shoot industries have expanded more rapidly. [The sector also includes the lesser (but important) flooring, mat, broom, and chopstick industries.] The sector has grown over the past 20 years, but its current rate of profit (8%) is approximately that observed in other industries in the county. This means that bamboo processing has no competitive edge in Longyou, and it is no longer a leading industry.

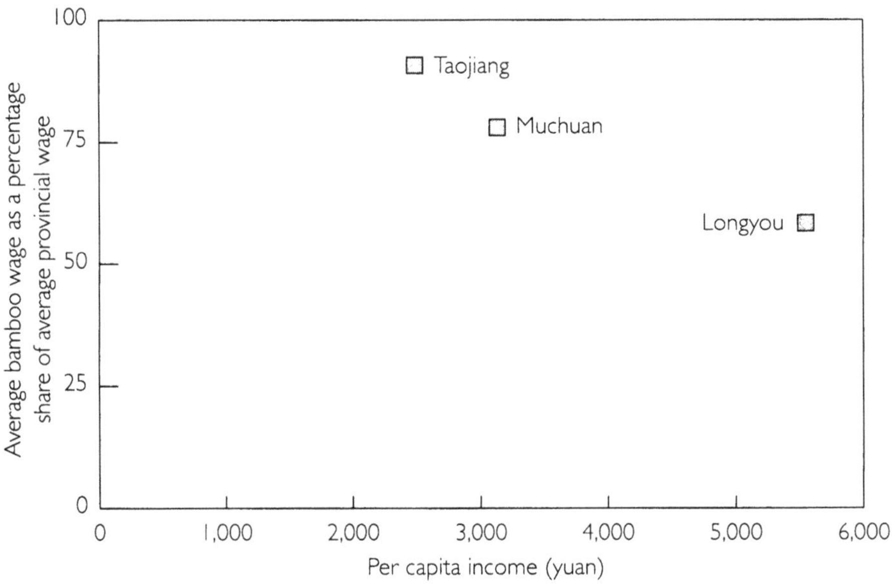

Figure 8-4. *Relationship between Per Capita Income and Relative Ratio of Bamboo Industry Wage, 1998*

Wages, Employment, and Farm Incomes

Wages and Employment in the Processing Sector. Forestry-sector wages were only 55% of the national average in 1998. Their relation to wages in other sectors varies widely. They were among the lowest of any sector in China. In our three counties, wages were 78% of the provincial average in Muchuan, 91% of the provincial average in Taojiang, and only 58% of the provincial average in Longyou.

The implications of these observations on relative wages become clearer when we compare them with per capita income in the forestry sector (Figure 8-4). They lead us to develop a three-part proposition:

- In regions where both the industrial diversity and the general state of economic development are greater, then industry in general and the forest industry in particular must compete for labor, and salaries and incomes in the sector are bid up. Forestry-sector salaries are higher than forestry-sector salaries in other regions because they have been bid up by the other local industries, but forestry-sector employment is still compensated at a lower rate than alternative employment in these regions. Because wages are high, this sector is not a likely candidate for the expansion of relatively labor-intensive forest activities.
- In regions where neither the industrial diversity nor the level of regional development is as great, employment alternatives are more limited, and forestry-sector wages are closer to those of alternative employment. All wages, including those for the forestry sector, are low—reflecting the lower level of

regional development and, perhaps, greater competition for low-wage employment.

- The opportunities for industrial expansion are greater in the latter case, especially for relatively more labor-intensive industries such as bamboo production and processing, for which low wages have the most beneficial effect on production costs.

The first case characterizes Longyou, which has a diversified industrial sector in general and a diversified bamboo-processing sector in particular. It is the most developed and has the highest per capita income of the three representative counties. Wages are higher in general in Longyou, but bamboo-processing wages are lower than those for other industries. The growth of the bamboo-processing industry in Longyou is probably past its peak.

The second case characterizes a county like Taojiang. Taojiang displays the greatest contribution of the primary bamboo-growing sector to county GDP, a low level of industrial diversity, and the lowest level of regional development of the three counties. Wages are low in Taojiang in general and in the bamboo mat industry that dominates bamboo processing. The bamboo processing industry is expanding more rapidly in this county than in either of the others because it takes advantage of the relatively low local wages.

Muchuan is an intermediate case with an established industry and a primary sector that makes an intermediate contribution to total county GDP. Yet, wages are still low, and the processing sector is still growing.

The evidence on women's employment complements our proposition. Muchuan and Taojiang employ smaller percentages of women in their bamboo processing industries, 46% and 47%, respectively. In contrast, women comprise 62% of the workforce for bamboo processing in more-developed Longyou. This difference is consistent with the argument that bamboo processing is a lower wage and otherwise less attractive source of employment in developed markets, and with observations from gender assessments in China that show women to be overrepresented in the less attractive sectors (Riley 1995, Hare 1999).

Farm Incomes. A perspective on the role of bamboo in economic development is emerging from our county-level observations, but what about the role of bamboo in the growth of household incomes and the alleviation of poverty? Does bamboo's contribution to household income form the same convex shape observed for Anji in Figures 8-1 and 8-2?

Farm household incomes increased in all three counties, much as they did for average households across all of China. They increased both most rapidly and in greatest absolute value in Longyou in the more developed eastern counties. Greater growth in the China's East and Southeast has been the common trend across all sectors and it has been a source of political concern.

Differences within each county's farm households are also growing. The Gini coefficients for households in our sample increased for all three counties. The smallest increase was from 0.24 in 1989–1990 to 0.25 in 1995–1996 for Longyou, while the largest was from 0.23 to 0.29 over the same period for Taojiang.[8] The increase in each county was largely due to an increasing disparity between incomes from farm and off-farm activities. Therefore, the increases in the Gini coefficients are largely due to activity outside the agricultural sector, and they probably are representative of the increased disparity between agriculture and those other sectors. They suggest that growing local income differences will reinforce the policy concerns with distributional issues that originally grew out of the pattern of regional differences between China's East and West.

Bamboo production and preprocessing in the home has generally improved farm household incomes in all three counties. The share of household income originating from bamboo activities increased without interruption for all two-year periods from 1989–1990 to 1995–1996 in all three counties.[9] Longyou households always earned the largest share from bamboo, 25% in 1989–1990 increasing to 41% in 1995–1996. Muchuan households experienced the most rapid gains, from 14% of household income in 1989–1990 to 37% in 1995–1996. Muchuan's expansion was due to the general expansion of the paper industry and new restrictions on the cultivation of steep land (more than 25° slope). It is consistent with our knowledge that bamboo production expanded more rapidly in the western counties. The expansion in Muchuan's infrastructure has been impressive, but more recent—and its effect on production is more recent as well.[10]

The distribution of bamboo income across households is increasingly diverse. Figure 8-5 shows how this distribution has changed for Taojiang. More households earned more income from bamboo in 1995–1996 than in 1989–1990. The spread in income levels also increased. Households in Muchuan and Longyou shared this experiences. The major differences across counties were the greater levels and wider spread of bamboo incomes in Longyou, two differences that we would expect in a more developed county with a more diversified economy.

Figure 8-6 shows the distribution of total farm incomes attributed to bamboo in 1995–1996, and Figure 8-7 shows the shares of the growth in income attributed to bamboo. Figure 8-6 compares with Figure 8-2 for Anji, and Figure 8-7 is related to the regression in Table 8-3. Once more, it is clear that bamboo contributed more to incomes in Longyou than in the other counties. It contributed least for households in Taojiang, at all income levels. Only the data for Muchuan demonstrate the convex form we observed for Anji, with middle-income households obtaining a larger share of their income from bamboo production. The lowest-income households never obtain large income shares from bamboo.

Figure 8-7 reinforces this second point. Bamboo is a more important source of income growth for high-income households than for medium- and low-income households in all three counties, and it was more important for middle- than low-income households in two of three counties. (An analysis of variance shows that these differences are statistically significant.)

These results support part of our findings in Anji. In particular, bamboo is not a "poor man's timber." On the contrary, in regions where bamboo has a leading role in economic development, the poor benefit, but they benefit proportionally less than those in higher income classes. In Anji, it was the middle-income and not the higher-income farmers that benefited most, whereas in the extended

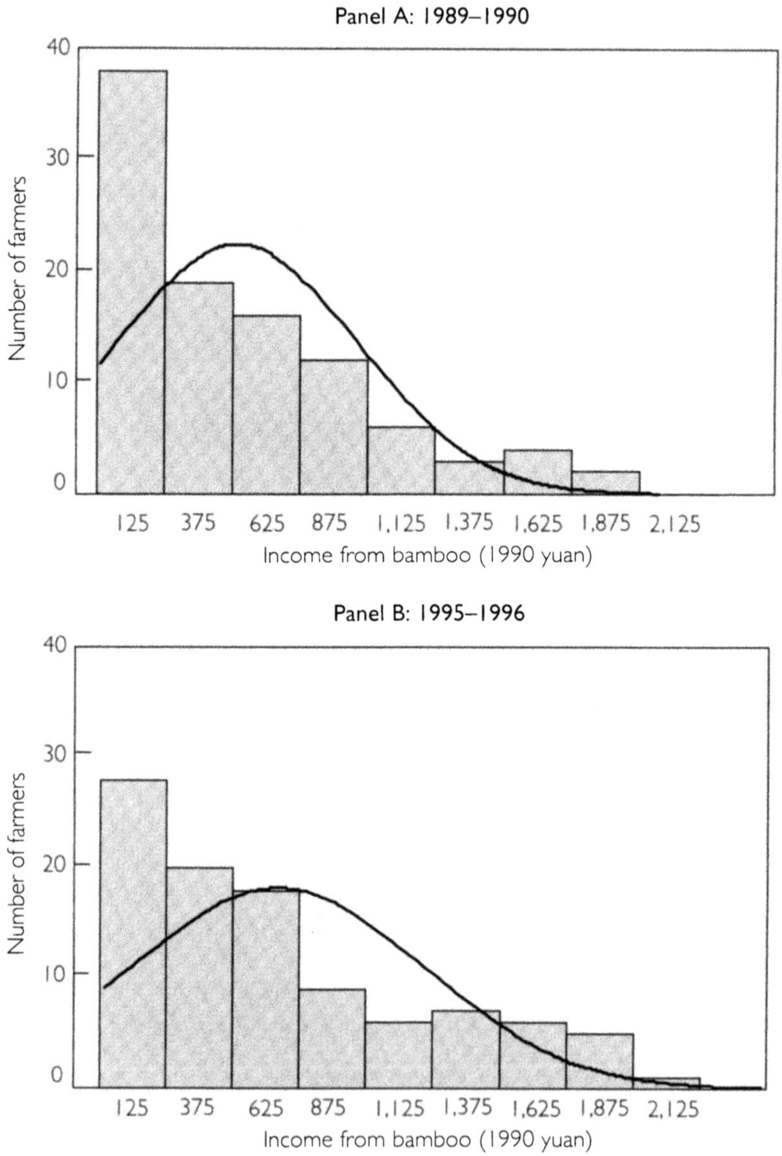

Figure 8-5. *Changing Distribution of Household Income from Bamboo in Taojiang*

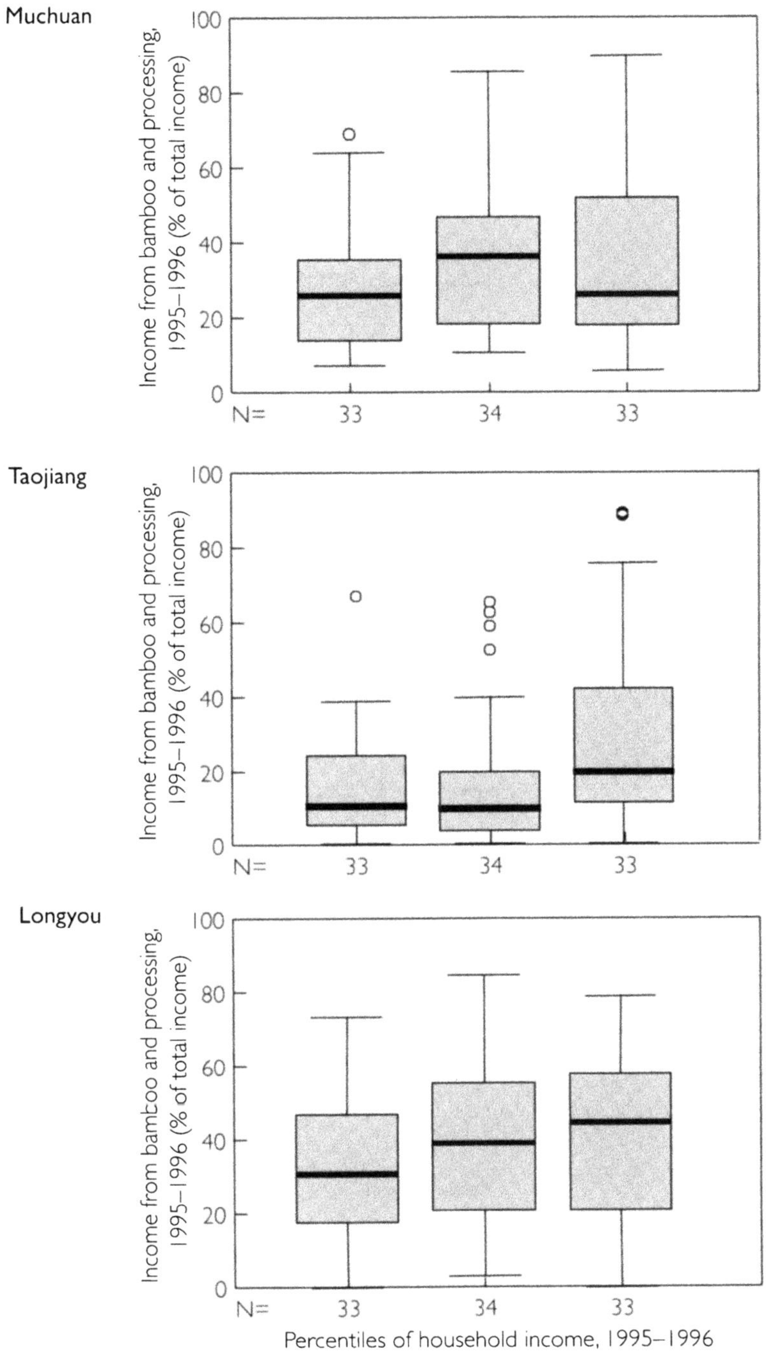

Figure 8-6. *Relative Importance of Bamboo to Household Income for Low-, Middle-, and High-Income Farmers in Three Counties, 1995–1996*

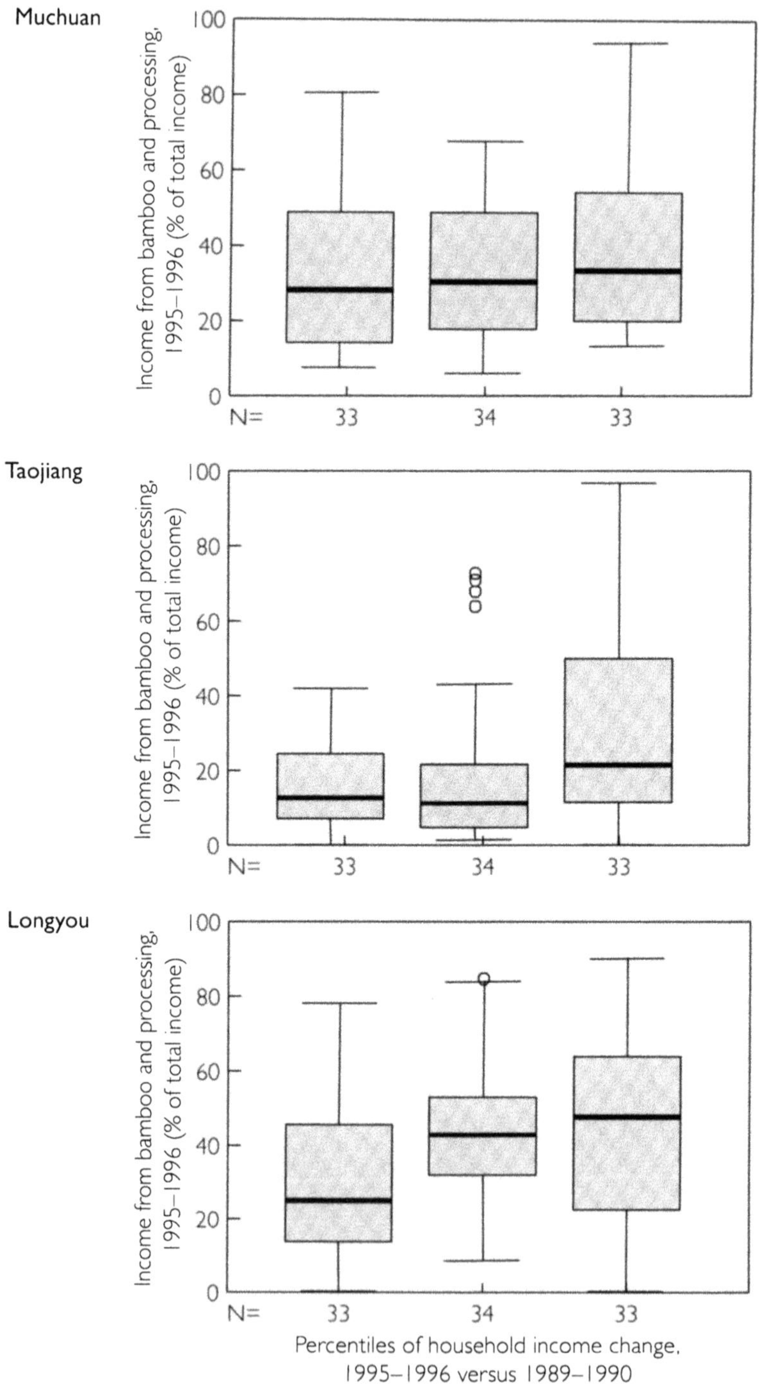

Figure 8-7. *Role of Bamboo in the Change in Household Income for Low-, Middle-, and High-Income Farmers in Three Counties*

sample of three counties, the higher-income farmers gained the greatest advantage.[11] [Middle income farmers in Anji occupy a place between those who do not need bamboo (i.e., the richer farmers who can earn more income from off-farm activities) and those who cannot take full advantage (i.e., the poorer farmers who do not have the resources to make the most of the opportunity).] However, Anji is far richer than the other five counties, with a per capita income that is twice as large as the next richest, Longyou. This causes us to offer the proposition that as the counties develop, their income distributions due to bamboo will look more like Anji's. The role of bamboo as a driving force for income growth will become less important, giving way to other more profitable activities (normally away from the farm) that will be taken up by more entrepreneurial farmers.

Conclusions

Can bamboo contribute to rural development and poverty alleviation? We have described China's experience with bamboo production and processing, with emphasis on the experiences of six counties. Bamboo frequently represents a superior option to traditional timber production. Bamboo forests, together with tree crops such as fruits and nuts, are classified as "economic forests" in China's official statistics. These economic forests are responsible for a major share of the expansion of forest investment that has occurred since reforms began in 1978. There are clear reasons for this, including earlier returns on investment (which, in turn, are related to pure financial considerations and to a wariness of policy inconsistencies), regular financial returns from annual or more frequent harvests, easier integration with agriculture (smaller units can be managed), and the absence of many regulations that act as disincentives for traditional forest investments.

This experience with bamboo is also instructive of the potential contribution of commercial timber production to poverty alleviation and rural development. If some of the taxes and regulations that act as disincentives for traditional forest investment were reduced, then timber production and the general performance of the timber sector could follow more closely many of our observations for bamboo. The markets are similar. On the demand side, bamboo and timber compete in many of the same markets—panels, flooring, plywood, fiber for paper production, and even some construction framing materials. On the supply side, they compete for similar farm resources because China's natural forests have been substantially depleted of their accessible bamboo and commercial timber. Therefore, expansion in either timber or bamboo production must largely come from planned investments in managed stands.

Farmers in the counties that we studied exhibited a proactive attitude toward bamboo, expanding their plantations when conditions were favorable. Bamboo processing was an important component of rural development. The small scale and low level of initial investment necessary in most bamboo-processing industries permitted the creation of a decentralized bamboo-processing industry, increased local demand and, therefore, increased local prices for bamboo. It offered farmers additional opportunities for income generation from investments of labor and capital in the farm, preprocessing activities on the farm, and off-farm bamboo employment in bamboo processing establishments. The improvements in local opportunity increased household wealth.

Bamboo's role in local enterprise and household welfare changes as the region develops. In the counties we studied, bamboo-based industries provide a good focal point for early industrialization, offering a number of opportunities, from markets for locally produced raw material to off-farm income and even some foreign currency. However, this pioneering role seems to be limited, and our evidence suggests that bamboo industries lose their dynamic nature as the county develops and enters more advanced stages of industrialization.

This does not imply that the bamboo sector can be dispensed with once development occurs. On the contrary, in Anji, the richest county in our sample, bamboo production and bamboo processing constitute a fundamental component of the county's industrial base, even if some other sectors now grow more quickly. It appears that bamboo will have a significant role in the county for a long time, especially as a supplier of raw material for Anji's modern, mechanized paper industry and as a producer of specialized bamboo products for the export market.

Policy reform has been instrumental in encouraging the development of the bamboo sector. Policy reform has changed the decision set, and the people have responded enthusiastically. The key elements of the reform have been the increased strength and security of land and resource tenure, the liberalization of international trade and investment policies allowing for private and joint-venture development and management of processing enterprises (with related gains in overseas investment and technology transfer), and the liberalization of domestic exchange allowing market mechanisms for the pricing and allocation of resources and providing easier access to domestic markets. These reforms have stimulated demand, and the resulting price increases have provided the incentive for intensified bamboo production. We anticipate that the same reforms would have similar effects in other components of the forestry sector as well.

What about the effects of bamboo-sector development on local income? Although most farmers in our county samples expanded their bamboo production activities and their bamboo-based income, the expansion was not uniform for all income classes. Middle- and higher-income classes of farmers did better, both absolutely and relatively. Bamboo preprocessing and off-farm activities in general offered the best opportunities for improved income, and the middle- and higher-income groups benefited most from these activities as well. Lower-income farmers seem to have benefited less, both absolutely and relatively. Apparently, market opportunities are not the most critical factors when it comes to explaining poverty. Improved markets and policies have improved the financial resources of all farmers, but many poor farmers remain relatively poor despite improvements in their incomes over the past 10–20 years.

As we conducted this research, we detected two trends that merit further inquiry. First, as development proceeds, we observe a trend toward specialization in some of the large family farms that are better positioned to take advantage of wasteland contracts, contracts for former collective lands, and other opportunities for land reallocation that could expand their forest-based activities. For example, we found farmers with 200 mu of tea—10 times the average holding of forestland in Longyou. Another farmer managed 700 mu of timberland and nearly 200 mu of bamboo land—20 times the average holding of forestland in Muchuan. The trend toward increasing scale and specialization also applies to the bamboo-processing industry, as in the case of a farmer in Taojiang who receives 115,000 yuan annually from his bamboo workshop. Interestingly, forest-based activities seem to be more suitable for expansion and specialization than agricultural crops, especially rice, at least at this stage of development and in the counties we studied.

This leads to the intriguing question of whether forest development is a source of increasing farmer segregation and differentiation in rural China.[12] The egalitarian base of rural China is changing as the economy develops and rural households exploit new opportunities. Farmers are becoming more specialized, and disparities in their income levels and their access to land and other resources are growing. Will a class of large landowners emerge? And what is the role of forest property for these landowners?

The second trend is toward an increasing specialization at the county level that links bamboo-based primary and secondary processing activities with tertiary activities. Anji County provides a prominent example with bamboo-based tourism.[13] Anji's position in the middle of one of the richest and most dynamic regions in China has provided incentives to shift from quantity to quality, including scenic values, landscape and species diversity, and a generally new environmental concern. New policies and accompanying economic measures to support them are shifting the focus of bamboo plantations from very intensive monocultures to more mixed systems in which maintaining the appropriate soil coverage and a larger number of species is a component of production (Ruiz Pérez et al. 2001).

Finally, we have focused on income and employment generation in the production and processing of forest products. There are many other important issues that we have been unable to address. For example, we have not discussed either the important subsistence and safety net functions of forests or the role of non-timber products from natural forests (a large subsector in China), or how either of these have been affected by policy reforms. Some policies have had substantial negative effects even as they may have improved the condition of the forest or the effectiveness of forest management. The massive unemployment created as subsidies for state forest farms have been reduced and the new logging ban is implemented is an example. The examination of these additional issues is part of our continuing research, and it will be an important element of Chapter 10.

Notes

1. Poverty is a multidimensional phenomenon. Although it is frequently characterized by reduced levels of consumption of food and material goods, poverty also has aspects of depriva-

tion related to lack of education, deficient health, vulnerability, voicelessness, and powerlessness (World Bank 2000).

2. Newer programs within the Western Regional Development Program and the Natural Forest Protection Program beginning in 1998 encourage reforestation and afforestation on degraded or sloping lands, especially in major watersheds.

3. Some marginal uses in drinks, medicinal applications, and even in the flavoring or coloring for rice do not fall within these two main categories.

4. The next two sections draw on the more extensive discussion of policy effects on the bamboo sector that appears in Ruiz Pérez et al. 1996.

5. This section draws on information previously published in more complete form in Ruiz Pérez et al. 1999.

6. Longyou was established as a separate county in 1983. Its county-level statistics begin with that year.

7. This section draws on Ruiz Pérez and Belcher 2001.

8. The *Gini coefficient* is a measure of income inequality: A coefficient of zero indicates that all households in a sample have the same income, whereas a coefficient of one indicates that all income is the possession of only one household. The Gini coefficient ranges from about 0.26 to 0.60 worldwide. A coefficient of 0.40 is normal for most developed countries.

9. For Anji, we examined bamboo income as a share of household income. In Muchuan, Taojiang, and Longyou, we examined bamboo income plus income from the preprocessing of bamboo products that occurs in the home. Preprocessing is important in Taojiang, less important in Muchuan and Longyou, and minimal in Anji.

10. Improved roads have cut travel time to Chengdu in half since the late 1990s.

11. The difference in percentage of income from bamboo between high- and middle-income farmers tends to be small, whereas the difference between these two income groups and low-income farmers is larger and statistically significant.

12. This process of segregation and differentiation takes place mainly through auctioning wastelands and other collective lands that include forests as an important component of their alternative uses (Hanstad and Li 1997).

13. Forest-based tourism is growing very rapidly and it has great potential in China. More than 63 million people visited forest areas (mostly in developed regions and highly populated areas around cities) for recreational reasons in 1995 (Da 1999).

References

Cao, Guangxia, and Lianmin Zhang. 1997. Preliminary Study of Rubber Plantation as a Local Alternative to Shifting Cultivation in Yunnan Province, China. Paper presented at the Indigenous Strategies for Intensification of Shifting Cultivation in Southeast Asia workshop. June 23–27, Bogor, Indonesia.

CFIC (China Forestry Information Center). 1998. *Development of China's Wood Industry.* Beijing, China: Chinese Academy of Forestry.

Chen, B., and J. Lu. 1999. China's Major Forest Regions: Natural Forest Protection and Sustainable Development. *Forestry Economics* (China) 5(2): 8–16.

China Statistical Yearbook. 2000. [In Chinese]. Beijing: China Statistical Press.

Da, F. 1999. Natural Forest Protection in Sichuan Province. *Forestry Economics* (China) 5(2): 37–43.

Gu, D. Q. 1992. The Influence of Bamboo Industry on Bamboo Growing Areas. In *Bamboo and Its Uses: International Symposium on Industrial Use of Bamboo,* edited by S. Zhu, W. Li, X.

Zhang, and Z. Wang. Beijing: International Tropical Timber Organization–Chinese Academy of Forestry, 329–332.

Hanstad, T., and P. Li. 1997. Land Reform in the People's Republic of China: Auctioning Rights to Wasteland. *Loyola of Los Angeles International & Comparative Law Journal* 19: 545–583.

Hare, D. 1999. Women's Economic Status in Rural China: Household Contributions to Male–Female Disparities in the Wage-Labor Market. *World Development* 27(6): 1011–1029.

Harkness, J. 1998. Recent Trends in Forestry and Conservation of Biodiversity in China. *The China Quarterly* 156: 911–934.

Jalan, J., and M. Ravallion. 1999. Are the Poor Less Well Insured? Evidence on Vulnerability to Income Risk in Rural China. *Journal of Development Economics* 58: 61–81.

Joint Survey Group. 1999. Sustainable Forestry Development and Achievements in Loess Plateau. *Forestry Economics* (China) 4(1): 10–20.

Li, S. and C. Xu. 1998. China's Bamboo Development Process and Development Strategies Towards the 21st Century. *Journal of Bamboo Research* 17(1): 1–5.

Li, Z., and G. Veeck. 1999. Forest Resource Use and Rural Poverty in China. *Forestry Economics* (China) 4(1): 80–92.

Liu, Y., W.C. Hsiao, and K. Eggleston. 1999. Equity in Health and Health Care: The Chinese Experience. *Social Science and Medicine* 49: 1349–1356.

MOF (Ministry of Forestry). 1995. *China Forestry Action Plan* [in Chinese]. Beijing: China Forestry Press.

Niu, W.-Y., and W.M. Harris. 1996. China: The Forecast of Its Environmental Situation in the 21st Century. *Journal of Environmental Management* 47: 101–114.

Peng, Zeyuan. 1999. Adjusting the Forest Benefit Structure to Push Forestry Sustainable Development. *Forestry Economics* (China) 5(2): 51–59.

Research Group of Forestry Economics. 1998. An Analysis of the Development Speed and Economic Benefit of Chinese Forestry Industry Enterprise. *Forestry Economics* (China) 3(1): 10–21.

Riley, N.E. 1995. Chinese Women's Lives: Rhetoric and Reality. *Asia Pacific Issues* 25: 1–8.

Ruiz Pérez, M., and B. Belcher. 2001. Comparison of Bamboo Production Systems in Six Counties in China. In *Proceedings of the Workshop on China Social Economics, Marketing and Policy of the Bamboo Sector,* edited by F. Maoyi, M. Ruiz Pérez and Y. Xiaosheng. Beijing: China Forestry Press, 18–54.

Ruiz Pérez, M., M. Fu, J. Xie, B. Belcher, and M. Zhong. 1996. Policy Change in China: The Effects on the Bamboo Sector in Anji County. *Journal of Forest Economics* 2(2): 149–176.

Ruiz Pérez, M., M. Zhong, B. Belcher, C. Xie, and M. Fu. 1999. The Role of Bamboo Plantations in Rural Development: The Case of Anji County, Zhejiang, China. *World Development* 27(1): 101–104.

Ruiz Pérez, M., M. Fu, X. Yang, and B. Belcher. 2001. Towards a More Environmentally Friendly Bamboo Forestry in China. *Journal of Forestry* 99(7): 14–20.

SFA (State Forestry Administration). 1999. National Report of the People's Republic of China: Towards ITTO Year 2000 Objective. *Forestry Economics* (China) 5(2): 85–102.

Smil, Vaclav. 1997. China Shoulders the Cost of the Environmental Change. *Environment* 39(6): 6–37.

Wang, C., and X. Zhao. 1999. Components and Measures for Integrated Mountain Development in China. *Forestry Economics* (China) 4(1): 21–29.

World Bank. 2000. *World Development Report, 2000–2001.* Washington, DC: World Bank.

Yu, G., L. Yu, C. Huang, and G. Chen. 1999. Survey and Exploitation on Forestland Transfer in Integrated Mountain Development. *Forestry Economics* (China) 4(1): 30–34.

Zhang, Z. 2000. Natural Forest Protection Program. First meeting of the China Council for International Cooperation on Environment and Development, Forests/Grasslands Task Force. Unpublished notes from presentations and discussions on July 18 and 19.

Zhang, Dongcheng and Rong Yuan. 1999. Problems and Countermeasures for the Development of Rosin Derivative Products for Naval Stores Industry in China. *Forestry Economics* (China) 4(1): 61–66.

Zhu, Shilin, Naixun Ma, and Maoyi Fu. 1994. *A Compendium of Chinese Bamboo.* Beijing: China Forestry Press.

CHAPTER 9

Impacts of Policy Reforms on Forest Environments and Biodiversity

Jeffrey A. Sayer and Changjin Sun

THE CHINESE LANDSCAPE contains an enormous range of forest environments, from tropical rainforests in the South and East, to alpine systems in the West, and deserts in the North. This great range of environments makes China one of the most biodiverse countries in the world, with 4,400 species of vertebrates (10% of the global total), including almost 500 mammalian, 1,189 avian, more than 320 reptilian, and 210 amphibian species. Its forests support approximately 8,000 species of woody plants, including 26 coniferous and 260 broadleaf genera. However, 15–20% of these species are threatened with extinction, and 200 species have already become extinct in recent years. Deforestation and the conversion of fragile lands for agricultural use are thought to be the most serious causes (Y. Zhang et al. 1999).

China's natural forests provide support for critical habitats of threatened and endangered species, forage and fodder for domestic livestock, fuelwood for the rural population, and herbs for food and traditional medicines that are in increasing global demand. They also support the rapidly expanding demand for forest-based recreation from a population that has increasing amounts of time and money to spend on this activity. China's forest parks absorbed 83 million visitors in 2001. In addition to these environmental services provided by China's natural forests, 17 million ha of recently afforested shelterbelts, predominantly in the North and Southwest, form the fundamental physical component of a vast effort to combat wind erosion and desertification, natural hazards that claim 2,000 km^2 each year (Gluckman 2000). Finally, the mountains and plateaus of western and northern China are the sources of five great rivers that flow into the world's most populated regions; the Yangtze, Yellow, Mekong, Heilong, and the Indus. The

The authors gratefully acknowledge the assistance of Tommy HuTao, Zhu Chunquan, and William F. Hyde.

water supplies for almost half of the world's population, living in China, Indochina, and South Asia, originate in China.

In sum, China's forests, including both its managed and natural components in both the uplands and the lowland river basins, are vitally important resources that contribute to both modern industries and rural household livelihoods. They provide environmental services of domestic and global importance, as well as commercial value.

China's forests have been subject to intense human use for millennia, but the human pressure on them has intensified in recent times. It was especially severe during the modernization of the early socialist period (1949–1978) and particularly during the Great Leap Forward (late 1950s). The loss and degradation of forests reached crisis proportions in the 1970s. In response, the Chinese authorities implemented several measures to promote reforestation and sustainable forest management. Chapters 1, 4, 5, 6, and 7 of this book discussed these measures and the forces that resulted in the expansion of China's production forests beginning in the mid-1980s and continuing even today. The increase in forest cover over the last 20 years has occurred at a time of unprecedented intensification of the pressures on the forest—from population growth, expanding economic activity, and a vast extension of China's roads and railways into remote regions. However, much of the increase in forest cover consists of "engineered" or managed forest and of orchards and gardens of commercial tree crops. On balance, the measures taken by the Chinese authorities to counter the loss and degradation of its forests have been widely proclaimed, but some have argued that success in increasing the land area under forest cover has been achieved at the expense of the natural forest and an associated loss of its biodiversity and other environmental functions (Smil 1993, Rozelle et al. 1997, Harkness 1998).

Our chapter will review these issues, beginning with summaries of the policies affecting forest-based environmental services and the recent history of protected area management. It will continue with a discussion of four forest resource goods or environmental services: protection for biodiversity, production of nontimber forest products, forest-based recreation and tourism, and watershed management and its impacts on downstream hydrological services. The general knowledge of each of these is limited, and our emphasis will be on biodiversity, where our knowledge is greatest. Our intention is to review the arguments regarding the gains and losses that have occurred in these goods and services and to reflect on the merits of each argument.

Background

Data

In China, as in most countries, the aggregate national statistics provide only limited information on the environmental values of forests. Even detailed measures of forest area and volume are insufficient indicators of nontimber forest values. For example, species-poor plantations composed of trees with poor soil conservation or hydrological properties may expand into areas of species-rich forests with high environmental value. Even degraded woodlands may have better soil- and water-conservation values and more indigenous biodiversity than the engineered forests that replace them. Aggregate figures on these environmental values of forests are notoriously difficult to assemble.

The Montreal Protocol and a plethora of other international attempts to develop criteria and indicators for measuring forest quality, demonstrate the global interest in this problem. The inadequacy of existing methodologies for assessing the environmental values of forests has been the force behind attempts to develop a forest capital index (WCFSD 1999), to implement a "millennium assessment" of the ability of the world's ecosystems to provide goods and environmental services [http://www.millenniumassessment.org/en (accessed July 21, 2003)], and to conduct multiple resource assessments to replace the national forest resource assessments that summarize a country's timber inventories (FAO 2001). All these initiatives face major methodological challenges and, as of 2001, only six countries had conducted even one multiple resource assessment. Timber remains the overwhelming focus even in these six cases.

In the absence of formal programs to monitor the environmental services of the forest, we must rely on China's published forest inventory data and the scattered experiences of knowledgeable observers. This will render our assessment inadequate—but it also leads us to one of our major conclusions: that more rigorous indicator sets are needed to evaluate forest quality. Therefore, we note the current efforts of China's State Forest Administration to work with international partners to develop criteria and indicators for sustainable forest management in two major regions, the temperate North and the tropical South. The need for these measures is especially urgent considering the large losses China is experiencing in its natural forest cover, the large investments in reforestation that are either planned or under way, and the considerable uncertainty that exists regarding the cost-effectiveness of these investments.

The discrepancies between the different data published about China's forest cover illustrate the difficulty of reaching firm conclusions about the forest condition. The definitive forest resource assessment (FAO 2001) is based on China's 1998 forest inventory. It reports that China has 163 million ha of forest and woodland (20.4% of its land area). It also reports that China's total forest area is growing by 1.806 million ha or 1.18% per year. However, these numbers are much larger than some earlier published figures. The 1998 State Environmental Protection Agency's country study on biodiversity (SEPA 1998, based on earlier FAO data and China's 1993 forest inventory) reports that China has a total forest area of 133.7 million ha (13.9% of its total land area). The discrepancy in the two measures of forest area occurs because the definition of forest cover that China developed for the 1998 inventory is broader than that used in earlier assessments. The requirement for the minimum crown cover necessary for inclusion in the forest stock decreased from 30% to 20% between the 1993 and the 1998 inventories. The change in definition means that some land in scrub forest and some newly planted lands that were not included in the 1993 inventory are included in the 1998 inventory. The latter—new plantations—are particularly significant because China has reforested so much so recently and because so many hectares of young growth have not yet achieved 30% crown cover. China's official data have never shown what the 1998 inventory would have been had the previous 20% measure been used continuously. Therefore, it is difficult to infer anything about changes in forest values from the official forest data.

Data from remote imagery may be an alternative to the official measures of forest cover and, in some cases, remote sensing imagery has been used to provide information about the extent of vegetative cover and the extent of changes in biomass. Remote imagery can allow the analyst to extract a measure of forest cover to satisfy almost any standard.

Runnstrom's (2000) assessment of coarse resolution satellite imagery for parts of northern China is an example. It provides independent confirmation that biomass increased between 1982 and 1993. The increase was partly due to the intensified management of agricultural lands and to tree planting for protection against the effects of flooding along the Yellow River. Whether these activities had any significance for environmental values like biodiversity is unclear. A second part of the increase in biomass was due to various measures taken to combat desertification. It seems likely that this second factor had a positive impact on biodiversity. Runnstrom's findings are significant in the light of past reports that many of China's afforestation claims were exaggerated and that the survival rate of planted trees has been low (e.g., Smil 1993).

Wider use of remote imagery in other regions of China as well could be beneficial. However, remote imagery is expensive, and it is clear that the measures of biomass and forest cover obtained from it may not correlate with biodiversity or the other important environmental values of forests. We are still left with imperfect measures of many forest-based environmental values.

Environmental Policy

Changing forest policy has had a large impact on China's forests and the biodiversity they contain. Three great periods of forest losses can be identified since the formation of modern China. First, forests were reallocated from the households for local management by the villages during the Great Leap Forward of the late 1950s. They were almost always destroyed quickly. Gangs of patriotic peasants scoured the country shouting slogans such as "Dig up all the things growing on the ground, shoot down all the things flying in the sky, dredge up all things living in water." The Cultural Revolution followed (1966–1976). Villages and communes were encouraged to smelt iron, and remaining forested areas were harvested to fuel the smelters. The third major period of forest loss occurred when the economy was opened in the 1980s. Rapid economic growth followed in all regions. During this period, timber farms exceeded their quotas and logged out many areas in the Northeast and the Southwest. Large-scale clearing of tropical forests in Xishuangbanna and Hainan also occurred to make way for rubber plantations. Many forest fires occurred, culminating in the fire in Heilongjiang in 1986. This single fire burned more than one million ha, one-third of which was destroyed beyond recovery.

Since then, tighter controls within the fire protection program and investments in fire warning and firefighting systems undoubtedly have reduced the incidence and damage of fires. These developments have probably been beneficial to biodiversity and to the other environmental services provided by forests.

Previous chapters have discussed the reform policies that led to a 16% increase in forest cover between 1978 and 1993 and, eventually, to a 9% increase in forest volume between China's third and fourth forest inventories in 1988 and 1993. (The new definition of forest cover applied to the fifth forest inventory in 1998 makes it more difficult to interpret recent changes in forest cover and volume.) However, these are aggregate improvements. They are the results of different policies affecting different lands. Therefore, they have a widely varying range of effects on biodiversity and especially on erosion control and watershed management. We will discuss them later in the chapter.

The legal basis for forest conservation has been greatly strengthened in recent years. A basic Forest Law was introduced in 1985, a Wildlife Protection Law was enacted in 1988, and several regulations for the implementation of these laws were passed subsequently. The central government took numerous measures to promote conservation in the 1990s. (The appendix to Chapter 1 outlines some of these.) The Natural Forest Protection Program, introduced in 1998, is notable among these. This program has a goal of protecting 61.1 million ha of forests in the upper reaches of the Yangtze and Yellow Rivers and 33 million ha in the Northeast and Inner Mongolia (CCICED 2000). Logging in the middle and upper reaches of the Yangtze and Yellow Rivers has been banned since 1998.

China adheres to the International Convention on Biological Diversity and recognizes in its National Biodiversity Action Plan that its forests are of global importance for biodiversity conservation. It has ratified several other international conservation conventions, including the Convention on International Trade in Endangered Species (CITES; since 1981) and the Ramsar Convention on Wetlands.

More recently, the central government has embarked on an ambitious program to return farmland on steep slopes (greater than 25°) to forest or grass cover. The goal was to convert 350,000 ha of these lands in the Yangtze and Yellow River Basins in the year 2000. Increments of 3 million ha more were to be afforested by 2003, and 5 million ha more by 2010. The plan is to return all cultivated land on slopes greater than 20° to forest or grassland within 30–50 years.

In addition, China's wildlife laws establish categories of protection for endangered or valuable species and impose penalties for killing or trading in banned species. Maximum penalties are very harsh, including long prison terms and even execution. More than 30 people have been executed for killing or trading in parts of elephants and giant pandas. In addition, China has added regulations to control imports and exports of all species listed on the CITES schedules.

In 1999, Premier Zhu Rongji identified sustainable development as one of two fundamental strategies for the twenty-first century (in the Government Report to the National People's Congress), and President Jiang Zemin stressed the importance of environmental protection (at the annual workshop on Population, Resource, and Environment). The president announced that any enterprise not in environmental compliance by 2000 would be closed. More than 2,000 small mills have been closed since then. Following these statements by China's most senior leaders, the Outline of the Tenth Five-Year Plan for National Economic and Social Development (2001–2005), adopted at the March 2001 annual meeting of the National People's Congress, places unprecedented emphasis on environmental protection and sustainable development. The targets for the five-year period include maintaining the human population below 1.33 billion people, halting environmental degradation, increasing forest cover to 18.2%, and greening 35% of urban areas.

In sum, China has in place a robust set of legal and programmatic initiatives aimed at the conservation of forest-based environmental services, and it has a strong commitment to these initiatives. However, these measures have been introduced at a time when the pressures on the forests have been intensifying. Population growth, economic expansion, and the penetration of new roads and railways into all parts of the country are increasing the demands for forest products and forestland. The decentralization of authority to local governments is increasing the pressures to expand economic activity at the local level, and this has not been countered with measures to protect the public good values of forests. The rest of this chapter will examine the extent to which the policy measures designed to conserve and increase forest resources match the increased pressures on these resources.

Protected Areas

The information on protected areas is clearer than the data on forest cover. Awareness of the need to set up nature reserves for biodiversity conservation has led to rapid and continuous growth in the extent of protected areas. China created its first national forest park, the Zhangjiajie Forest National Park, in 1982. Many of the earliest reserves were set up specifically for conservation of the national treasure, the giant panda, but these were quickly followed by other reserves established for their scenic, wetland, or ecological values or to conserve other species. By the end of 1999, 1,118 nature reserves had been established on 86,410,000 ha (8.6% of the national territory). The eventual target is to protect 10% of all land in nature reserves (BWG 1997, SEPA 1998, FAO 2001, Harkness 1998). All levels of government participate in the management of these reserves.

Most of the protected areas in southern and eastern China are small because very little suitable land remains and human population densities are very high. In contrast, some reserves in western China are very large. However, they consist of deserts and mountains and contain little forest and little biodiversity. For example, Fajin Shan Reserve in Xinjiang covers 45,800 km^2, an area just larger than Switzerland. Changtang Reserve in Xizang is several times larger, with a total area of 247,120 km^2.

The expansion of the land area in nature reserves might be expected to yield significant benefits for biodiversity and for ecotourism and watershed management. The information on the management of these protected areas is limited and anecdotal, but the evidence that does exist indicates that the quality of management is generally poor, but highly variable (MacKinnon et al. 1996, SEPA 1998).

One major problem is that the finances allocated for the development of nature reserves are usually once-off investments. Adequate funding is not generally provided for their continuing operation. Therefore, the work units responsible for each reserve are encouraged to develop their own economic ventures to generate the funds to support the operation and maintenance of the reserve. As a result, many commercial ventures have been established within the protected areas: zoos, amusement parks, breeding to produce endangered animals for sale, sales of mounted specimens of rare species, illegal harvest and cultivation of timber, poaching of wildlife, and uncontrolled development of ecotourism. All of these ventures bring financial resources into the system and, thereby, help finance the objectives of protected area management. However, many of them also conflict with the declared objectives of protected areas.

For example, the Wolong Nature Reserve and the Giant Panda Breeding Station in Sichuan have received consistently high levels of financial support for many years. Nonetheless, small towns, dams, and a market containing restaurants and souvenir shops have been established inside the reserve. The Wuyishan Nature Reserve in Fujian is another example. Wuyishan is one of China's better-developed nature reserves. It benefits from the investments of a relatively wealthy province and from the support of the World Bank's Global Environmental Fund. Nevertheless, the management staff at Wuyishan has converted land inside the reserve into plantations of moso bamboo (*Phyllostachys heterocycla*), which is harvested and manufactured into chopsticks and other products by small factories run by the authorities of the reserve. These enterprises are financially successful and they do allow the reserve to continue operation without further external financial assistance, but they also diminish the value of the reserve for biodiversity. The bamboo plantations effectively fragment the reserve into several isolated forest blocks and create barriers between the upland summer ranges and lowland winter ranges of many of the altitudinal migrant species that live in the reserve. Studies of the bird communities in the reserve in 1993 showed that the bamboo plantations supported fewer species than nearby villages, farmlands, or tea plantations (personal communication from John R. MacKinnon).

A second problem is that the infrastructure and staff of the nature reserves are generally located in towns that are some kilometers away from the natural areas themselves, and the staff do not visit and patrol the reserves adequately. Although the signs of illegal activities, cutting, poaching, and mining are apparent in many protected areas, few transgressors are arrested and brought to court. The personnel of the reserves generally lack the power to arrest, and they do not have the resources to arrest and detain transgressors or to support the costs of prosecution.

Biodiversity

China has suffered serious losses in its biodiversity. The generally accepted estimate for the global loss of species is 10%, but the estimate for China is a much greater 15–20% (Chen 1993). The establishment and maintenance of protected natural areas is one step in improving this condition—although we have indicated that this step has its own problems. The effect on biodiversity of the last quarter century of forest policy is debatable, and China's new openness to international trade has been less beneficial.

Trade in Wild Plants and Animals

The opening of the economy in the 1980s led to profound changes in the patterns of consumption of wildlife and native plants in southern China. As China's more affluent population was able to afford exotic foods, farmers invested more time collecting wild plants and animals for sale. In a few cases, they began to grow plants and animals that were no longer available in the wild.

As the traditional resources were exploited and depleted, imports became more important and the volume of imports of endangered plants and animals increased enormously. The increased trade in wildlife was most apparent in border provinces. Monitoring at two posts along the Vietnam border in Guangxi Province revealed that an average of more than 14 tons of wildlife passed each post each day in mid-1994. The estimated total annual value of the wildlife trade at these two posts alone was more than 500 million yuan (US$58 million), and more than 10 such posts operated along China's border with Vietnam in 1994 (Li and Li 1997). The range of species involved was broad—including assorted snakes, monitor lizards, geckos, freshwater turtles, monkeys, lorises, civets, and pangolins. Many CITES-listed species were included even though China had joined CITES in 1981 and Vietnam had joined in 1993.

Trade in the Hong Kong markets also increased for a wide range of species. For example, imports of chelonians (turtles) increased dramatically until 1994: from 139,200 kg in 1977 and still about the same level at 110,574 kg in 1991, to 680,582 kg by 1993 and 1,800,024 kg in the first 10 months of 1994. Market monitors in Hong Kong observed trade in a total of 375 avian species, 13 mammalian species, 79 amphibian species, and 278 reptilian species (Lau et al. 1997).

Melville (1982) showed that the export of songbirds—mainly Peking robin, hwamei, mesia, magpie robin, and white eyes—through Hong Kong reached the level of about half a million birds a year by the late 1970s and early 1980s. Subsequent investigation (Nash 1993) showed a reduction to 200,000–300,000 per year by 1990. However, with many more export channels now available, total exports may now be much greater. Formerly common birds such as Peking robin and hwamei have been added to the CITES annexes. The registered trade, especially in songbirds, is massive, and illegal trade compounds the problem. Falcons smuggled from northwestern China to the Middle East and the *shatoosh* wool of the Chiru antelope smuggled to India are two such examples.

These wildlife are consumed for food and for the production of traditional Chinese medicines (TCMs). TCMs are widely used throughout China, side by side with modern pharmaceuticals, and they have become a multi-billion-dollar export industry to Chinese communities overseas. Guo et al. (1997) examined the use of wildlife for TCMs. They visited factories or mailed questionnaires in order to identify species and quantities. Of the 17 most widely used wildlife products, all except worms, rat snake, ground beetle, pearl, pilose antler, centipede, and toad venom were no longer in adequate supply by the mid-1980s. Many others (e.g., pangolin, musk deer, bear, leopard, monkeys, sea horses, and gecko) are now endangered in China. Supplies of bear gall, musk deer gall, and pilose antler can only be met by farming, and other species are so rare that substitutes are increasingly used in medicinal preparations. The situation is similar for plants. Orchids are stripped off branches for the production of cough medicines; alpine meadows are stripped of fritillary bulbs, orchid tubers, gentians, poppies, and other important plants.

The opening of the domestic market has led to the massive capture of Chinese wildlife for export. It has also led to large-scale external criticism of China's growing list of internationally threatened species. The criticism caused the government to ban the sale of tiger and rhino parts in all medicines in 1993—at a cost to the industry of US$240 million during the first year of the ban. China also hosted a special international workshop in 1995 to discuss wildlife trade issues. Twenty neighboring countries attended, and all agreed to tighten the controls on cross-border trade. The agreement had an immediate effect on biodiversity. Teams monitoring the border posts in 1995 found that the volume of wildlife trade declined to about one-quarter of its volume in the previous year. In addition, protected species were less openly available in the restaurants and wildlife markets of southern China.

Meanwhile, China's expanding international trade has led to another problem in the form of invasive alien species. Both accidental and deliberate introduction are at fault. For example, each giant grain ship entering the port of Shanghai carries within its hold an unintended cargo of about 20 tons alien weed seeds. Meanwhile, exotic grasses are intentionally imported for the erosion control of newly constructed road embankments. Pantropical weeds such as water hyacinth and *Eupatorium* are becoming pest weeds in the southern provinces.

The Effects of Forestry and Forest Management

Between the second forest inventory in 1978–1981 and the fourth in 1991–1993, China afforested 6 million ha of shelterbelts, 5 million ha of orchards and other tree crops, and 2 million ha of nature reserves and fuelwood forests. China's foresters also afforested 14 million ha in new managed ("engineered") forest plantations, some on converted wasteland, and some on former cropland or pasture. Finally, China reforested 9 million ha that had been and remained forest or that substituted for land that had been forest. Some of the reforested 9 million ha was clearcut, then reforested. More was selectively harvested, then reforested naturally or by planting around the remaining natural stand. An unidentified amount of this 9 million ha was new forest that offset land that was cleared of forest cover elsewhere. At the same time, China lost 9 million ha of its remaining natural forest. The net impact was an impressive gain of 18 million ha in total (managed plus natural) forest cover. (See the calculations by Rozelle et al. in Chapter 6.)

The 9-million-ha decrease in natural forest is an undeniable loss in native habitat. Beyond this point, there is some disagreement on the impact on wildlife habitat and biodiversity. The introduction to the World Wide Fund for Nature–China (WWF–China) annual report for the year 2000 points to the return of bird life not seen in 25 years to the reforested lands within a few hours drive of Beijing (Harkness 2001). Surely, some wildlife population is also recovering in other newly afforested areas.

Some observers are more concerned with the losses of biodiversity associated with losses of old-growth forest and with clearcuts of natural forests that are replaced with monoculture forest plantations (Smil 1993, Rozelle et al. 1997, Harkness 1998). Our experience does not support these concerns. First, we have observed very few harvests of old growth and very few clearcuts in our visits to China's forests. In fact, very few old-growth forests remain in China. Most of China's old growth was cut long ago. The forests we observe now are almost all secondary forests—often naturally grown, but secondary forests nonetheless. Second, selective harvests are much more common than clearcuts when the objective has been the removal of commercial timber. In fact, high-grading has been the most common practice for timber harvesting throughout China. High-grading removes the best trees and leaves the lowest value trees and species standing. Professional foresters detest high-grading because it leaves the worst trees as a basis for future production, but high-grading also leaves the sort of dog-eared gnarly old trees and commercially undesirable species that create good habitat for a diversity of wildlife. Third, many of the new forests are the result of replanting with seedlings of one, two, or three species. Sometimes the new seedlings are from introduced species that were not native to the area, but usually these are planted among the remaining high-graded stand, or else they are planted on open land with no remaining standing forest. In the former case, the advantages of high-grading for biodiversity remain; in the latter case, even these forests tend to support a variety of wildlife in their understories and canopies as they mature. The forests referred to as supporting new birdlife in WWF–China's 2001 annual report were forests of the latter sort.

Net Losses of Natural Forests. Harvesting and reforestation tend occur in two different locations. There are exceptions, but this has generally been the case in China, and it certainly has been the case since the beginning of reforms in 1978.[1] Therefore, the loss of natural forest cover probably was significantly greater than the 9 million ha referred to above, and this loss probably justifies our greatest concern. It could be as great as 9 million plus the 7 million that just offset harvests, for a total of 16 million ha. Even the 9 million is an immense number, 7.9% of China's total forest cover in 1978 and an unknown but much larger share of its remaining natural forest. This 7.9% (or more) was lost in only 15 years!

The replacement of the natural forest with forest plantations is not the same as reintroducing natural forests at a new site, but that cannot really be done anyway. New plantations today grow into a commercial resource sometime in the future. Therefore, they create a resource that substitutes for future harvests from the natural forest. Today's plantations actually will help reduce the future draw on the natural forest and its biodiversity, and the future forest environment would be even less diverse without these new resources.

Perhaps a better way to look at the changes in China's forest cover would be to recognize that the drawdown of natural capital like forests is a well-recognized component of economic development. Furthermore, after some level of development, most countries tend to improve their natural environments, including their forests. Thus, the greatest potential losses in natural habitat and biodiversity occur in the early stages of development, and there is a special need to identify and protect critical habitat and the most threatened categories of forest environment during this period. This is a primary justification for developing rigorous indicator sets of critical habitat and sustainable forestry. For China, it means that the current losses in natural forest cover may be greater than they will be in the future. In some ways that would be fortunate. But it also means that the preservation of biodiversity deserves greater attention now because it is also more difficult now than it will be in the future. (The appendix to this chapter provides a classic example of this sequence for the case of Sichuan Province.)

Increases in Forest Cover. Losses in natural forest are a real problem. Gains in new forest cover are not, but they cause us to inquire about the importance of a new problem. More than two-thirds of the expansion in China's forest cover occurred as the result of reforestation of the collective lands by individual agricultural households. (Follow the gains in collective forestland in Table 1-2 in Chapter 1.) These new forests replaced crops, pastures, and sometimes wasteland. Farmers do not tend to manage their forests as intensively as they manage their agricultural lands. Therefore, these new forests probably support a greater diversity of life now than in their former uses as cropland or for grazing livestock.

Nevertheless, in our own travels across the agricultural countryside, we notice considerably less birdlife in China than in many other countries. The new collective forests should support increasing amounts of wildlife. Do they support a greater diversity of life than before? If they do not, then are the new forests less diverse than the same lands were in their previous uses? Alternatively, does the demand for wildlife trade cause farmers to harvest whatever wildlife does remain? Or, as a third alternative, do China's more productive and more prosperous farmers now use more pesticides on their crops than before? We don't know the answers, but the question is essential for the long-run well-being of China's environment.

Nontimber Forest Products

Surely China's reforms, and its tremendous economic growth, have had an effect on the nontimber products of the forest. However, the data for these products are just as problematic as the data for biodiversity. Much of the consumption of these products takes place outside the market, and the nontimber forest goods that reach the market are pooled with similar or identical goods grown by some farmers. Therefore, the best we can do is to reflect in a general way on three categories of these goods: specialty products such as herbs, mushrooms, and TCMs; fuelwood; and forage and fodder for domestic livestock.

The demands for specialty foods and TCMs have increased with the increasing wealth of Chinese consumers and with the opening of China to international markets for these products. In some cases, as in bamboo (discussed by Ruiz Pérez et al. in Chapter 8), farmers have begun growing more, even as they continue to extract from the natural forest. For some herbs and TCMs, most of the supply still comes from the declining natural forest. The net effect is unknown. In fact, whether the consumption of TCMs has a serious impact on the forest and what will happen as the natural supplies decline is a matter of conjecture. Can farmers begin to grow many of these products, or will consumers learn to substitute for them or do without? What will the impact be on the natural forest? We cannot even speculate.

It is easier to speculate on fuelwood because there is widespread evidence from many consumers in many economies around the globe. In general, low-income rural households have inelastic demands for fuelwood. However, once their incomes rise sufficiently, they begin substituting liquefied petroleum gas, kerosene, and other processed fuels, and the aggregate demand for fuelwood becomes elastic. Furthermore, as supplies from the natural forest become scarcer, rural landholders begin growing fuelwood for themselves and for the market. What these patterns mean for China's forests and for its rural population is, once more, conjecture. It would be useful to know the points at which (a) China's regional markets will begin to substitute alternative fuels in a significant way and (b) farmers will begin growing their own fuelwood. Surely, these points will differ between broad regions in a country as vast as China, but how much forest will be removed before the substitution of homegrown fuelwood begins and whether a change in consumption or production will cause hardship for any part of China's population are not questions that are within either our expertise or our knowledge of China's forestry or agricultural literature.

The demands placed on the forests and woodlands by China's domestic livestock are different. Once more, rising household incomes have increased the demand for meat and the demands for forage and fodder for livestock—and once more, the impact on the forest is unknown. However, we can identify two important policy effects.

First, China's food security policy restricted imports of meat and replaced them with domestic production. This restriction was not important for pork, for which domestic production always has been sufficient. It was important for the rising demands for beef and mutton. It meant restricting low-cost imports of these products from central Asia and feeding more domestic livestock on the economically marginal and environmentally risky grasslands and scrub forests of western China. How great were the environmental losses? Once more, this is conjecture. There are anecdotal accounts of rodenticides being used to control pikas and other rodents over very large areas on the premise that these small animals are responsible for some of the observed overexploitation of grasslands. However, the burrowing activities of these animals almost certainly play a role in maintaining the fertility and productivity of grasslands. It is more likely that grassland degradation can be attributed to overgrazing by domestic livestock.

Second, neither the Household Responsibility System contracts for cropland nor Contract Responsibility System contracts for forests included the common grazing lands. The use rights for the common lands went unchanged in the 1980s and 1990s. Therefore, the common lands did not benefit from the improved household incentives and the longer-term conservation investments introduced by individual farm households on other croplands and forests. They continued in a degraded state until the new Western Regional Development Program in 1998 began permitting the transfer of some of common pastures and scrub forests to local households, and the Natural Forest Protection Program prohibited grazing on the steepest slopes and provided incentives for their reforestation. The latter are large-scale programs. It will be important to observe the effect they have on the old common grazing lands in the years to come.

Forest-Based Leisure Activities and Environmental Tourism

Forest-based leisure activities and environmental tourism are among the benefits obtained from the protected areas we already discussed. In general, these activities grow rapidly with increasing personal income, and that has been China's experience as well. The general tourism data for China show 777,710 person-visits in 1982 (*China Statistical Yearbook* 2002) and 83 million person-visits for forest parks alone in 2001 (SFA 2002). By 2001, these visits to forest parks contributed 37.3 billion yuan of gross revenue (an increase of 25% from 2000) and supported 3.5 million jobs.

This is explosive growth and it probably will continue, although perhaps at a slower rate, as China's household incomes continue to grow. With growth in the use of the forest parks, however, new environmental problems have arisen. Some protected areas are beginning to suffer from pollution, congestion, and other environmental damage. The control of these negative effects of growth will become an increasingly important problem.

Watershed Management and the Hydrological Functions of the Rivers

China suffers from some of the most serious erosion problems of any country in the world. It loses 2,000 km^2 of arable land annually to wind erosion and desertification. In response, it has embarked on a massive afforestation effort, including more than six million ha of shelterbelts planted between 1979 and 1993, and many more hectares planted by individual farmers as components of agroforestry systems. The well-known Three North Forest Protection Project (Green Great Wall) in North Central China is part of this effort. The survival rate of seedlings planted on many of the afforested hectares is often questioned, but there can be little doubt about the net positive environmental effect of this effort (Runnstrom 2000).

China's river courses also suffer from the effects of erosion. In their case, it is the upstream erosion that settles downstream as sediment in the lakes and river courses themselves and on their floodplains. Logging and road building in the uplands are among the causes. They alter the rates of water flow and, with them, the natural rates, levels, and locations of siltation.

The lower Yellow River is of particular concern. Fifty years ago, 50 lakes were connected to the lower river. Complex patterns of river flow allowed fish to migrate between the lakes and the main river. In some seasons, water flowed from the flooding river into the lakes; in others, the flow reversed. The region was rich in fish species, and it produced valuable harvests, especially of the four domestic species of carp (grass carp, *Ctenopharyogodon idellus*; silver carp, *Hypophthalmichtys molitrix*; bighead carp, *Aristichtys nobilis*; and black carp, *Mylopharyngodon piceus*). At one time the region provided two-thirds of the nation's freshwater fish harvest. Today, some fish species are extinct, several are endangered (nationally 92 species are listed as endangered, as are wildlife such as the Yangtze alligator and freshwater dolphin), and the fish catch has declined dramatically. The two largest lakes, Dongting and Poyang, have been reduced to almost half of their former surface area and only one-fourth of their original depth. Other lakes have been separated from the main river by the construction of sluice gates and the reclamation of land for agricultural crops. Siltation has also has caused a reduction in water transparency and a loss of macrophyte vegetation, with a consequent loss of herbivorous fish. Data from Lake Honghu show a reduction of annual fish catch from 10,000 tons in the 1950s to only 3,000–4,000 tons by the 1980s. Catch from Lake Dongting decreased from 45,000 tons/year to 15,000 tons/year over the same period. Lake Honghu had been the habitat for 90 species of fish. This number declined to 74 by 1964 and to 54 by 1982—of which only 33 species actually lived in the lake itself (Xie and Chen 1997).

In addition to these effects of logging, road building, and downstream siltation on the freshwater system, some of the river's increased sediment load reaches the sea where it smothers the southern coral reefs and reduces the catch from the inshore saltwater fishery. Finally, plans to block the Yellow River at Three Gorges and divert water to northern China will have undetermined additional effects on both the downstream freshwater systems and also the inshore saltwater fishery.

China's rapidly growing wood-processing industry is another source of serious environmental intrusion. The pulp and paper industry has grown at a double-digit annual rate since 1982. It has become the source of 10% of China's industrial wastewater emissions and one-fourth of its chemical oxygen demand. It is the largest source of rural environmental pollution (Huang and Bai 1992, State Council 1996). The industry's most rapid growth has occurred among the thousands of small township and village enterprises (TVEs), which also are the most damaging to the environment. The pollution control technologies used in the pulp and paper industry are designed for large (capacity greater than 100,000 tons) wood-fiber-using mills, but the annual capacity of almost all of the TVE mills is less than 10,000 tons, and these mills rely on nonwood fibers at more than a three-to-one ratio over wood. The government shut down more than 1,000 of these mills for noncompliance with environmental standards in the late 1990s (Xu 2000).

China's entry into the World Trade Organization and the anticipated influx of capital that it will bring may shift China's paper production toward larger capacity, less-intrusive, mills, and imported wood may sustain them. However, the logging ban associated with the new Natural Forest Protection Program will probably force the very many remaining small inland mills to rely on even greater shares of agricultural residues and other nonwood fibers. Nonwood fiber processes yield more solid waste effluents than wood fiber processes—to the further detriment of China's watercourses. Therefore, although the logging ban in the uplands may protect the upland environment and improve the hydrological functions of China's rivers, its effect on the paper industry may be more detrimental, causing a serious unanticipated negative effect on the downstream environment. Clearly, there are serious trade-offs to consider, and the best means to address them are not at all obvious.

Conclusions and Future Prospects

As in many economies going through periods of rapid development, many members of Chinese society and many components of the Chinese economy and its natural environment have experienced substantial gains during the past 25 years. However, the forest environment has suffered serious losses, and it continues to suffer serious losses. It is crucial to the long-run well-being of the country that the sources of these losses and the losses themselves are addressed.

The government embarked upon the massive Western Regional Development Program (WRDP) in 1999. The WRDP has two basic objectives. It intends to bring the level of economic development of the interior and western parts of China up to the levels already achieved in coastal and eastern parts of the country, and it intends to protect the upland environment from further deterioration. The second objective contains three elements that are important for the forest environment: forest closure, reforestation of bare hills, and the restoration of steep farmland to forest or grass cover. All three are designed to increase the area under forest cover and, thereby, improve the forest environment, and also to decrease erosion in the upper watersheds and limit downstream sedimentation. They will require large investments from the Chinese government and international donors, as well as significant sacrifices of labor and land by local farmers.

The full implementation of the WRDP in all of its aspects is still unclear, and its effects will not be known for some time. It remains to be seen whether the program can control either the losses of natural forest and its diverse biota or the serious effects of upland economic activity on the downstream hydrological function of China's rivers. The WRDP is a bold action and surely will have benefits. It warrants close observation for what we can learn from it and also because we cannot anticipate that it will be a complete solution for all forest-based environmental problems everywhere in China. Some old problems (perhaps desertification and the loss of habitat and species in lowland forests) will remain, and some (such as pollution from the paper industry) may be intensified.

The most important lesson to be learned from China's experience has to do with political decisionmaking regarding the forest. The most drastic losses of China's natural forest occurred during the political movements known as the Great Leap Forward and the Cultural Revolution. That is, political reasons instead of economic development (commercial logging) have been the primary source of the loss of natural forests. Activities such as clearing mountains for grain production and terracing simply did not and do not make much financial or economic sense. They did not produce much grain at all, and the land areas involved have largely returned to forest, although not to their original vegetation patterns. Two large programs being implemented today as part of the WRDP—the logging ban on natural forests and the conversion of sloping croplands to forests—are reversing some of the detrimental results of past political initiatives. While the direction of change is correct, the cost is high and, unfortunately, the original vegetation cannot be restored at many of the deforested sites at any cost. We can hope that the world will learn from China's experience, and that other countries will not repeat the politically driven mistakes that China has made in the past.

Appendix: Sichuan's Experience

Until the 1960s, forest cover in the middle catchments of the Yellow River (*Changjiang*) was about 45% of total land area, and timber harvest levels remained modest. However, as China's economy developed, the exploitation of the forest accelerated. Timber harvest levels in Sichuan rose from 100,000 m^3/year in the 1950s to 3,000,000 m^3/year in the 1960s to 5,000,000 m^3/year in the 1980s.

After 1980, timber from the conversion of forestland to agriculture augmented the production from forest farms. Farmers harvested still more timber for fuel and for house construction. Overall forest cover declined to as little as 16% of total land area by 1985. The decline in forest cover greatly damaged the hydrological functions of the river system, and the severe fragmentation of the remaining forest was a threat to biodiversity. Forty percent of the province's habitat for the giant panda was destroyed during this period. Pressures from hunting and the collection of medicinal plants and other forest products increased, and the forest area continued to shrink.

Timber harvest practices and forest management techniques have changed since the 1980s, and major efforts have been made to reforest bare slopes. The area under forest cover has increased to about 24% of the provincial land base. However, much of the new forest is immature and, therefore, it will be awhile before it provides much protection against erosion and accelerated runoff.

The floods of 1998 and 1999 were blamed for both the loss of the forest sponge effect for the siltation of downstream drainages. The central government responded by banning logging in the upper reaches of the Yellow River watershed. It has also implemented a major program for restoring forest cover for the purposes of restoring hydrological functions. It plans to restore forest cover to 28.4% of total land area in the province by 2005.

Note

1. The replacement of the natural forest with rubber plantations in Hainan is probably the best known exception. It occurred in the 1950s.

References

BWG (Biodiversity Working Group). 1997. *Conserving China's Biodiversity: Report of BWG to the China Council for International Cooperation on Environment and Development (CCICED).* Beijing: CCICED.

CCICED (China Council for International Cooperation on Environment and Development). 2000. Report of the First Meeting of the CCICED, Forest/Grasslands Task Force, July 18–21, 2000. Beijing.

Chen, Lingzhi. 1993. *China's Biodiversity: Current Strategy and Its Conservation Strategy.* Beijing: Science Press.

China Statistical Yearbook. 2002. [In Chinese]. Beijing: China Statistical Press.

FAO (Food and Agriculture Organization of the United Nations). 2001. *Global Forest Resources Assessment 2000.* Forestry paper 140. Rome: FAO.

Gluckman, R. 2000. The Desert Storm. *Asiaweek,* October 13, 36–40.

Guo, Yinfeng, Xeying Zou, Yan Chen, Di Wang, and Sung Wang. 1997. Sustainability of Wildlife in Traditional Chinese Medicine. In *Conserving China's Biodiversity: Report of the Biodiversity Working Group to the China Council for International Cooperation on Environment and Development (CCICED).* Beijing: CCICED, 190–216.

Harkness, J. 1998. Recent Trends in Forestry and Conservation of Biodiversity in China. *The China Quarterly* 156: 911–934.

Harkness, Jim. 2001. Introduction. In *WWF China Annual Report 2000.* Beijing: World Wide Fund for Nature.

Huang, Qinan, and Donghai Bai. 1992. *Development Strategy of China's Paper Industry.* Beijing, China: China Light Industry Publishing House.

Lau, M.W.N., G. Ades, N. Goodyer, and F. Zou. 1997. Wildlife Trade in Southern China Including Hong Kong and Macao. In *Conserving China's Biodiversity: Report of the Biodiversity Working Group to the China Council for International Cooperation on Environment and Development (CCICED).* Beijing: CCICED, 141–159.

Li, Yiming, and Dianmo Li. 1997. Status and Strategies for Control of Live Wildlife Trade Across the Sino-Vietnamese Border. In *Conserving China's Biodiversity: Report of the Biodiversity Working Group to the China Council for International Cooperation on Environment and Development (CCICED).* Beijing: CCICED, 128–140.

MacKinnon, John, Meng Sha, Catherine Cheung, Geoff Carey, Zhu Xiang, and David Melville. 1996. *A Biodiversity Review of China.* Hong Kong: World Wide Fund for Nature.

Melville, D.S. 1982. A Preliminary Survey of the Bird Trade in Hong Kong. *The Hong Kong Bird Report* 1980: 55–102.

Nash, S.V. 1993. *Sold for a Song: The Trade in Southeast Asian Non-CITES Birds.* Cambridge, U.K.: Traffic International.

Rozelle, Scott, V. Bezinger, Jikun Huang, and Li Guo. 1997. The Rise of Forests and the Fall of Diversity: Assessing the Role of Tenure, Policy and Markets. Paper presented at MacArthur Foundation Workshop on Biodiversity in the Himalayan Region. April 7–10, 1997, Kunming, China.

Runnstrom, M.C. 2000. Is Northern China Winning the Battle Against Desertification? *Ambio* 29(8): 468–476.

SEPA (State Environmental Protection Agency). 1998. *China's Biodiversity: A Country Study.* Beijing: China Environmental Sciences Press.

Smil, Vaclav. 1993. China's Environmental Crisis: An Inquiry into the Limits of National Development. New York: M.E. Sharpe.

State Council. 1996. Resolution on Several Issues in Environmental Protection [in Chinese]. Unpublished document no. 37.

SFA (State Forest Administration). 2002. Press conference release for First China Forest Landscape Resource Exhibition and Tianmu Mountain Forest Festival. June.

WCFSD (World Commission on Forests and Sustainable Development). 1999. Final Report. Cambridge, U.K.: Cambridge University Press.

Xie, Ping, and Yiyu Chen. 1997. Biodiversity Problems in Freshwater Ecosystems in China: Impact of Human Activities and Loss of Biodiversity. In *Conserving China's Biodiversity: Report of the Biodiversity Working Group to the China Council for International Cooperation on Environment and Development (CCICED).* Beijing: CCICED, 160–168.

Xu, J. 2000. China's Paper Industry: Growth and Environmental Policy during Economic Reform. Ph.D. thesis. Blacksburg: Virginia Polytechnic Institute and State University.

Zhang, Y., G. Dai, H. Huang, F. Kong, Z. Tian, X. Wang, and L. Zhang. 1999. The Forest Sector in China: Towards a Market Economy. In *World Forests, Society, and Environment,* Vol. 1, edited by Matti Palo and Jussi Uusivuori. Dordrecht, the Netherlands: Kluwer Academic Publishers, 371–393.

CHAPTER 10

Conclusions and Policy Implications

William F. Hyde, Jintao Xu, Brian Belcher,
Runsheng Yin, and Jinlong Liu

CHINA HAS UNDERGONE a remarkable transition since the introduction of agricultural reforms in 1978. The summary effects are well known: 20 years of average annual growth in excess of 10% and a fourfold increase in per capita income. Industrial, financial, and then trade reforms followed the initial agricultural reforms, and nearly all members of Chinese society have benefited.[1]

The effects of these reforms on China's forests and the forest sector of the economy, and the effects of the specific forestry reforms themselves, provide a rare perspective on many questions of modern forest policy—for China and for the rest of the world. Furthermore, this perspective comes at a moment when global concerns about greenhouse gases, biodiversity, and the preservation of endangered species, as well as more regional concerns with water supply, erosion control, the general environmental services provided by forested watersheds, and rural economic well-being, have enhanced everyone's awareness of forest policy.

The lessons from China's reforms are important as China considers the best programs for sustaining its remarkable growth. They are also important for the rest of the world. China alone can have an immense impact on the world's forests. China's population, one-fourth of the world's total, has reduced its own forest cover to 0.10 ha per capita, a ranking that is among the lowest for any country in the world, and China has begun to rely on others for its wood products. Its roundwood imports nearly tripled, from 4.8 million m^3 in 1999 to 13.6 million m^3 in 2001 (*China Customs Office Statistical Yearbook* 1999 through 2001).

China's lessons are also instructive for policy design in the rest of the world. Many eastern European and some Asian countries are undergoing a similar economic transition from centrally planned to more market-oriented economies. Others, such as Indonesia and the Philippines, are making the transition from economies with highly concentrated political and market power to a wider distribution of economic authority and less-administered markets. Countries such as India are attempting broad market deregulation. Each can learn from China's experience.

China's experience is even more broadly relevant yet, because roughly half of China's forests, like those of most developing and developed countries, have been under the authority of a central forest ministry. China, like most countries today, faces the difficulties of centralized forest management in the presence of powerful local values. In fact, the decentralization of forest management and the development of management regimes that are more responsive to local economic and social pressures may be the most widely discussed topic of modern forest policy. Some degree of decentralization is known as "community participation" in the United States, Canada, and Finland, as "joint forest management" in India, as "collaborative forestry" in Uganda, and as "community forestry" or "community-based resource management" in Bolivia, Colombia, Mexico, Nepal, the Philippines, Tanzania, Uganda, and Zambia. Similar terms are used for related experiences in a host of other countries.

The common question has to do with just how far a government can go with decentralization and even privatization of its forests while still satisfying the broader public values that forests provide. One widespread argument against decentralization is that farmers have little experience planting trees and, furthermore, they would not plant trees even if they had the experience. Their time horizons are just too short. China now has more than 20 years of experience against which to test this fundamental perception. A related argument holds that individual farmers are not responsive to the natural environmental benefits of reforestation because those benefits are shared widely. This argument anticipates that the benefits accrue to the farmer's neighbors and even to those who live farther away and who cannot exercise the persuasive power of one's neighbors. Therefore, it concludes that private action is unlikely and that some level of public action will be necessary to ensure these environmental services. China has experience against which to examine this argument as well. Indeed, China has had more experience than any other country in the world in transferring some centralized forest administration and some forest land-use rights to local administration and even to individual households.

Policymakers and policy analysts debate the effects of agricultural policy and overall economic growth on the forest. China's experience is instructive on these points as well. Do rising agricultural prices and agricultural support policies destroy forest, as they may have in Cote d'Ivoire (Lopez 1998) and Belize (Chomitz and Gray 1996), or is agricultural development more likely to have a favorable impact on forest cover as it has in India's Punjab (Singh 1994)? Does economic reform and general economic growth improve the condition of the forest? The sparse analytical evidence on these questions is mixed (Jepson et al. 2001, Pandey and Wheeler 2001). China's experience covers a greater range of growth than most developing country experiences for which we have data. It can add perspective that is unavailable elsewhere.

In fact, this leads us to a final point. China's lessons are often convincing because the evidence is so very clear. China's great geographic breadth, its tremendous economic growth, and the pragmatic and experimental nature of its reforms (called "crossing the river one stone at a time") mean that the data incorporate great regional and intertemporal variability—and variability in the data allows for both evidence from contrasting examples and statistically meaningful results. As a result, the same lessons that may hold anywhere in the world are often clearer within the context of an experience like China's.

This final chapter provides an opportunity to review the basic insights emerging from China's experience. They can be summarized within six crucial issues of modern forest policy:

- property rights and improved land tenure, especially farmers' behavior with respect to trees and forests;
- property rights and alternative institutions—the choices among state forest management, community management, and private forestry;
- general policy effects on forests—including uncertainty in policy implementation, agricultural policy spillover effects onto the forest, and the impacts of the overall macroeconomic policy environment;
- forestry's potential as an agent of economic development;
- the environmental impacts of extractive commercial activity in the forest; and
- forestry data.

The first five lead us to some generally overlooked questions about the last issue, forestry data, and about the appropriate analytical distinctions for meaningful inquiry into forest policy. These insights are important for forest policy globally. Some of them are especially important for China's own future.

We will close with a brief discussion of the emerging issues for China's forestry in the twenty-first century—and the measures its government must take if wishes to continue the development of the forestry sector and the improvement of China's forest environment.

Private Property Rights and Improved Land Tenure

Prior to 1978, China's agricultural production was centrally organized by collectives and production teams. Even in China's most productive northern plains, annual productivity was as low as 800 kg of grain per ha, and the base of arable land was declining. Per capita annual income in the collectives was only 63 yuan (approximately US$25) (*China Agricultural Yearbook* 1985, Zhong et al. 1991). Meanwhile, China's forests were managed in roughly equal proportions by the collectives and the state-owned forest bureaus, although the bureaus were increasing their share as they gradually encroached on the best collective forestlands. Forests covered 8.6% of China's total land area, and most forests were seriously depleted. Standing forest volume on the state lands was in the neighborhood of 100 m^3/ha, while the collective forests averaged less than 50 m^3/ha—in comparison with the global average of 100 m^3/ha (FAO 2001).

Agricultural reforms began locally in Anhui Province with household responsibility system (HRS) contracts that transferred agricultural land-use rights from the collectives to individual households. HRS contracts were relatively short term, usually three to 15 years in duration. Households were also given the right to sell their production in excess of assigned quotas at market prices. These reforms were attractive to government officials as a means of reducing their transactions costs and the risks of centralized management. The reforms also improved local household incentives and productivity. They spread rapidly, even before receiving the official sanction of the central government in September 1980. By 1984 HRS contracts had expanded to 70% of all rural communities. Agricultural productivity exploded. Productivity in the northern plains, for example, increased 7.5-fold to 6,000 kg/ha by 1984. One element of the reforms, the two-price system, was difficult to maintain and eventually, by 1985, farmers were permitted to sell their entire production at market prices.[2]

Forest-dependent households observed the success of the agricultural reforms and demanded similar reforms in their own sector. A contract responsibility system (CRS) similar to HRS spread to forest households in the Southern Collective Forest Region in 1981—as the third component of what was known as the "Three-Fix Policy" (stabilizing the rights and ownership of forests and mountains, identifying boundaries to household plots, and establishing a forest production responsibility system). It too spread rapidly. The share of collective households participating in CRS increased to 55% of those in forest collectives by 1983 and, by 1984, more than 30 million ha of collective forest had been designated for private management by 57 million households (*China Forestry Yearbook* 1987). (Fewer than 20 million ha remained under centralized collective management. The state-owned forest bureaus remained responsible for almost 50 million ha.) Forestry-sector reforms, however, were not as complete as agricultural reforms. Forest households were required to sell their production to monopsonistic state wood-processing operations. In fact, state procurement remained a constraint on market activity in forestry in many regions, even after this marketing policy was liberalized in 1985.

Despite the limitation on timber markets, the effects on forest cover were rapid and substantial—on both agricultural and forest lands. Since agricultural reforms occurred first, we expect to observe the first effects on those agricultural lands. Yin, in Chapter 4, observed that, in the agricultural North Central Plains, forest cover expanded from about 5% in 1977 to 11% by 1988. The expansion was most apparent in shelterbelts, in intercroppings of trees with agricultural crops, and along the "four sides" of villages, houses, roads, and waterways. Farmers planted trees in these locations as a means of protecting against the region's severe wind erosion. Yin used an agricultural production function to assess the effects of improved property rights (under HRS), market liberalization, and the conservation investment in trees, as well as the standard agricultural inputs, on agricultural output. He concluded that improved property rights provided the stimulus for almost 60% of the increase in agricultural output between 1978 and 1984. Conservation investments in trees (a response to improved property rights) explained 5% of the increase in output through 1984 and 20% thereafter, as newly planted seedlings matured into larger trees and formed better protection against eroding and desiccating winds.

The northern plains, however, may be an atypical example because agriculture was so important in that region that the authorities paid little attention to forests. Liu and Edmunds, in Chapter 2, examined the experience of southern China, a contrasting region where forests were much more important. Both agricultural and forestry reforms began in the South, and this region contains about 85% of China's collective forestlands. Liu and Edmunds reviewed the various institutional arrangements for land transfers, and they cautiously reminded us of the multiple restrictions on full transfers of rights. They also observed that standing timber volume declined in eight of 10 southern provinces between 1977 and 1984. Standing volume declined as well in almost all of the individual villages for which Liu and Edmunds had more specific evidence. However, the land area under forest cover increased over the same period, probably by more than 20%.[3] Both the standing volume and the area of household-managed forests in southern China have increased in every official forest survey since then (*China Forestry Yearbook* 1989, 1998).

Zhang, in Chapter 5, pointed out that individual households now manage the bulk of the former collective lands in all regions of the country. As household responsibility increased, the area of collective land under forest cover and its standing volume increased simultaneously. Nationwide, between 1988 and 1998, collective forest cover expanded almost 40% to 67 million ha, while the standing forest inventory on the collective forestlands grew almost 35% to 2,962 million m^3. Volume grew less rapidly than forest cover because some of the new forest cover was still composed of juvenile growth, which is not easily measured in the official forest inventories, and some of it was composed of fruit trees that are included in China's official inventory but which never accumulate large volumes.

In sum, the evidence is compelling. Farmers do plant and manage trees. Once farm households received limited land-use rights, they expanded the area under forest cover and they increased the standing forest volume in every region of the country and for all three official forest inventories since 1977. Furthermore, the increases occurred despite the uncertainties of changing contractual rights to the land and changes in other local policies and administration; despite harvest quotas, shipment restrictions, and monopsonistic buying that limited market opportunity; and despite increasing forest taxes. We will review some of these restrictions later, as they point to measures that China can take if it seeks to continue the growth of its forestry sector and the improvement of its forest environment.

This first lesson of China's farmers is crucial. Farmers and other small landowners will plant and manage forests, once they obtain the rights to the land and its forest resources and if trees and forests are the highest valued uses of their lands.

Property Rights and Alternative Institutions: Community Rights and State Management

A second question—and a second lesson—follows naturally from the first lesson about private investment. That second lesson has to do with the appropriate institutional level for forest management. State natural resource management agencies and local community organizations receive most of the attention in current discussions around the world.

The general arguments for state management are that private individuals do not possess a sufficiently long-term perspective for forest management and that the externalities and broader social values inherent in forests require management by an institution that is responsive to the views of all of society. However, the extensive evidence of illegal logging and other encroachment on public lands around the world, the inefficiencies of and even gross mismanagement by some large public agencies, and increasing recognition of legitimate local claims on resources may recommend more local collective management. In fact, community management ("community forestry") seems to be the preference of many nongovernmental organizations and many forestry advisors in the large development agencies.

China's experience rejects all the easy answers. It provides evidence of successful private management and rejects the contention that private forest management does not have a sufficiently long-term perspective. However, China's experience with upstream erosion and the downstream flooding of the vast Yangtze River Basin in 1998 and with recurrent drought in the Yellow River Basin, experiences that led to the new Natural Forest Protection Program and a ban on logging in the upper watersheds, also seems to reject any contention that private or even local community management is always sufficient. Finally, the failures of both China's state-owned forest enterprises (SOFEs) and its forest collectives argue that neither state management nor local collective action is always sufficient either. Eighty percent of the SOFEs have exhausted their mature timber resources and cannot support their own mills. Fifty percent of them were operating at financial losses by 1997, and two-thirds owe back salaries to their workers (Li 1996, *China Forestry Yearbook* 1998, Zhang 2001). The increases in China's forest cover and standing forest volume that occurred over the last 15–20 years were due to private investment, and they occurred only after state authority over some collective lands was rescinded and the collectives transferred land and forests to private households.

Therefore, the evidence from China is insufficient to suggest a "best" arrangement for the management of many forest values, whether market or nonmarket, whether local, regional, or even global. However, China's experience does offer two contrasting cautions on the limits of both community forestry and private management as the only ways to provide for some local values.

- Private management provided a level of forest protection against the nonmarket effects of soil erosion in China's northern plains. Collective action was unnecessary, at least for lower levels of protection, in this severely erosive but highly productive agricultural region.
- Although private household management has worked well, economies of scale have made collective action more successful for some accumulations of the smallest plots of forestland. (See Chapter 2.) China's egalitarian distribution of former collective forestland resulted in the fragmentation of forested plots. Households held as many as four or five plots, some as small as 0.06 ha, combined to average less than 0.6 ha per household. Boundaries per unit of forestland were longer than they would be with larger plots and the costs of bound-

ary marking and forest protection from fire and theft were greater than some households could afford. This led some to see the advantages of consolidation, and transfers and consolidation did occur in some regions. In others, notably Fujian, Guizhou, and Zhejiang Provinces, the problems with managing fragmented holdings led to self-initiated shareholding systems (Liu 2001). These allowed for land consolidation, shorter boundaries, lower protection costs, and better collective access to capital for forest investment, while building on local capacity and maintaining local incentives for forest management (Yin 1998).[4]

The conclusion to be taken from China's experience is that the efficient level of administrative control is dependent on both the scale economies of management and the particular social values at hand. Universal prescriptions, whether of private management, collective community forestry, or state enterprise, are ill advised.

General Policy Effects on Forests

China's experience provides clear examples of the effects of policy uncertainty on longer-term investments like those in forestry. It also provides an opportunity to examine the effects of market and policy spillovers to forestry from other sectors and from general macroeconomic activity.

We are all aware that many developing countries have experienced sharp and rapid swings in land tenure, agricultural support, and trade policy, as well as distinct shifts in fiscal and monetary policy, government administration, and even civil tranquility. We know the fiscal balance has had an impact on the standing forest in countries like Brazil and Indonesia (Young 1996, Jepson et al. 2001), and Deacon (1994) has shown that general administration and civil unrest are important determinants of global deforestation. Yet the forest policy literature generally concentrates on the selective effects of particular forest markets and policies and largely disregards longer-term policy uncertainties and those spillovers from other sectors and the macroeconomic policy environment. We need more evidence of their importance, and China's experience provides it.

Uncertainty in Forest Policy and Administration

Yin, in Chapter 4, and Zhang, in Chapter 5, provided complementary insights to the effect of an uncertain forest policy environment.

Frequent policy changes, even before the initial agricultural and forestry reforms, had made cautious investors of most households. National policy on the ownership and management of forestlands changed four times in the quarter century before 1978, and local administration often changed even more frequently. Farmers were doubtful that any policy would be long-standing. Therefore, when they obtained CRS rights to forests in the early 1980s, they often harvested what they could almost immediately—before the policy could change again.

Yin examined the impacts of this policy uncertainty on timber harvests and forest cover by contrasting the behavior of households in a region of greater policy certainty (the North Central Plains) with that of households in a region subject to frequent shifts in policy and administration (the Southern Collective Forest Region, or SCFR). Forestry was less important to the northern plains economy, and local administrators did not impose large tax or regulatory burdens on farm forest activities. Households in this region rapidly developed a level of confidence in the new tenurial arrangements, and they planted trees. Their harvest levels rose initially upon receipt of the new CRS rights, and they continued to increase over time because investments in forestry increased and, therefore, both new seedlings and the standing timber resource grew.

In some parts of the SCFR, where forests are a larger part of the economy, local forestry administrators determined that the initial round of private timber harvests meant that households could not manage trees and forests. They withdrew CRS rights from some households. This only increased the uncertainty for those households that still held CRS rights. They promptly increased their timber harvests and reduced their investments in forest management. Standing forest inventories declined and, as a result, so did forest growth and subsequent timber harvests.

With additional time, the policies and administration governing land and forest tenures stabilized, and many farmers in the SCFR eventually increased their investments in forestry.[5] Both forest cover and forest volume increased in the SCFR after 1988. However, most household forest managers retain a degree of uncertainty today, because the local availability of harvest and shipping permits for timber changes from season to season, because forest taxes and charges have risen from negligible to 35–60% of timber revenues, and because some local forest administrators seem to prefer a level of arbitrariness in their oversight—particularly for the more successful forest managers (Chapters 2 and 3).

Zhang, in Chapter 5, used more recent evidence from China's emerging stock markets to examine the effects of these continuing uncertainties on China's larger private forest operations. He observed that equity in forestry firms trading in the late 1990s displayed an industrywide level of risk that was an immense 45% greater than the level of risk displayed across all firms trading in the same markets. This means the financial return had to be 45% greater in forestry in order to induce investments equal to those in other industries.[6] As long as the risks causing this investment behavior prevail, China can anticipate a low level of private investment in forests and forestry.

Spillover Effects from Other Sectors

The second question about broader effects on the forest sector has to do with spillover effects from activities in other, nonforest, sectors of the economy. Rozelle et al., in Chapter 6, and Y. Zhang et al., in Chapter 7, conducted econometric examinations of the full range of forest and nonforest market and policy effects on forest cover.[7] Both sets of authors included independent variables measuring changes in forest tenure, agricultural markets, and the general economy. Y. Zhang et al. also incorporated terms for both general uncertainty and uncertainty that was specific to the application of forest policy. Their findings for these uncertainty factors were consistent with those discussed previously. That is, the presence of uncertainty decreased the level of forest investment and the subsequent forest stock.

Rozelle et al. observed that general economic growth was an important source of the increase in China's forest cover. Demands for the amenity services of forests have expanded rapidly with China's more recent economic growth, just as the demand for fuelwood has declined with emigration from the rural areas in search of urban employment and as households have converted from fuelwood to natural gas and coal as their incomes have increased. The net effect of the expanding demands for both industrial wood and amenity values of the forest, together with a declining demand for fuelwood, has been an increase in total forest cover.

Rozelle et al. observed only a small and insignificant effect of relative timber:agricultural prices on the standing forest, probably because agriculture can have offsetting effects on managed and natural forests. That is, increasing agricultural prices induce agricultural competition with, and expansion into the area of, managed forests. The agricultural price effect on natural forests is less predictable, but increasing prices could induce a concentration of extensive agricultural activities that previously took place in natural forests, thereby decreasing agriculture's competition with natural forests and increasing the land area in those forests.

In Chapter 7, Zhang et al. improved on the analytical approach by distinguishing between managed and natural stands and between general agricultural prices and the prices of those crops grown on marginal agricultural lands, the crops that are most likely to compete with forests for land use. They made two important sets of observations. First, increasing prices of competitive agricultural crops (rubber and horticultural crops in the tropical province of Hainan, their region of inquiry) induced land conversion out of managed forests, but saved natural forests by inducing intensive rubber and fruit production rather than extensive extraction of these products from the natural forest. Second, managed forest cover expanded in response to growth in the general population and the economy, as well as to increasing wood prices and improved land tenure.[8] The two general economic factors (population and the aggregate economy) were responsible for 25% of the increase in managed forest cover over the period from 1957 to 1994. Agricultural prices and improved tenure were responsible for 58% of the increase. The timber market itself was responsible for only 15% of the increase.

Summary and Discussion

First, general policy uncertainty and changes in the administration of China's forest policies have had a substantial negative effect on investments in forestry and on total standing forest volume and forest cover. We can anticipate that the avoidance of large swings in the orientation of forest policy and the evenhanded administration of the prevailing policy will be central to the future success of long-term forest management in China. China's experience should convince us of the importance of these two issues elsewhere in the world as well.

Second, the two examinations of broader policy spillovers onto the forestry sector are instructive—if not universally definitive. Both show that general economic growth goes hand in hand with significant improvements in the level of forest cover. Moreover, Zhang et al. reminded us that economic growth tends to coincide with stronger institutions and, therefore, a reduction in the transactions costs of forest management. This is a further reason to anticipate a relationship between economic growth and improved conditions for forestry. Nevertheless, few examinations of the general economy's impacts on the forest sector have been conducted for any country anywhere. We need confirming evidence from other parts of the world, but this evidence from China suggests the hypothesis that economic growth and development leads to increases in forest cover and provides a positive response to concerns for global deforestation.

Forestry's Potential as an Agent of Economic Development

Questions about forestry's role in economic development have been with us at least since the emergence of the large post–World War II development agencies. The questions arise because extensive forest cover often occurs together with poverty and isolated human communities. Initially, the simple existence of forests was seen as a means to support an industrial base built from sawmills and pulp and paper facilities (Westoby 1987). Subsequently, we learned the advantages of forestry's multiple forward and backward linkages with other economic activities, as well as the advantage of the great range of alternative technologies and, therefore, substitution possibilities between capital and labor in logging and wood processing (Hirschman 1978). Despite these advantages, forestry has not demonstrated great success as a development agent, and more recent attention has turned to forestry as a supplementary source of support for the rural poor (e.g., Wunder 2001).

Can forest development help poor households escape poverty and, for China, what might the pattern of forest development look like if some of the current constraints on forest development—such as high taxes, government procurement, and restrictions on timber harvests and shipment—were removed? The pattern of development in China's bamboo sector since 1980 suggests answers to both questions.

Bamboo and commercial timber are similar in their productive inputs. Both use rural labor and land, and neither is a significant competitor for agricultural capital. Bamboo also competes in many of the same product markets as timber, providing raw material substitutes for timber in paper and various construction activities. Bamboo production has grown more rapidly than commercial wood production—from 3.2 million ha in 1980 to 4.3 million ha in 1999, and from 4.4 million metric tons of output in 1980 to 14.2 million metric tons in 1999 (Ruiz Pérez et al. 2001). Bamboo's more rapid growth is partly due to the absence of harvest quotas and shipment restrictions and to the much lower schedule of taxes and fees for it than for timber. Taken together, these similarities to and differences from timber mean that bamboo's pattern of development is suggestive of what the pattern for commercial timber would have been under more liberal market conditions. This contention was reinforced when China introduced a logging ban over large areas of the country in 1998. Bamboo prices immediately rose 5–10%, and the land area in bamboo plantations grew nearly 17% in one year (CFIC 1998).

Ruiz Pérez et al., in Chapter 8, reviewed the bamboo experience for six counties along an east–west gradient from Anji Province on the Pacific coast to more remote inland parts of Hunan and Sichuan Provinces. They observed that improving bamboo markets provided an alternative income source for many farm households and an opportunity for some households to diversify and specialize. They grouped farm households into five income classes and observed that households in all income groups gained from rising bamboo prices and market opportunities, but middle- and higher-income groups gained relatively more. Bamboo opportunities were important, but off-farm employment opportunities were even more important sources of increasing household income, and middle- and higher-income households obtained more of these opportunities as well.

Bamboo production increased, but neither the number of farm households nor the land area devoted to bamboo production changed substantially in most counties. Rather, farmers tended to increase production by intensifying management on the same lands.

An expanding bamboo-processing industry, located close to its raw material supply, stimulated demand and provided new opportunities for rural labor in both contracted home-based preprocessing activities and in the factories themselves. Policy reforms external to the forestry sector formed a critical basis for this development by permitting the local small industries (known in China as township and village enterprises), including bamboo-processing enterprises, to market finished products in excess of their production quotas and by facilitating the export of their finished products (bypassing inefficient state marketing boards). Subsequent reforms in the 1990s made private domestic investment more attractive and permitted foreign investment, including investment in the bamboo-processing industry. The collective effects of these reforms illustrate the importance of forward linkages to bamboo production. They also reinforce the arguments of the previous section of this chapter regarding the importance of spillover effects from other sectors onto production within the bamboo and forestry sectors themselves.

Those counties that continued to develop eventually diversified their economic base. More off-farm employment opportunities developed, and wages rose, as did the opportunity costs of land. The financial returns to bamboo production tapered off. Farmers who remained active in bamboo production diversified, specialized, and intensified their productive activities in order to remain competitive in their use of land and their demands for labor.

In sum, bamboo production was a good starting point for early industrialization, particularly for counties with good access to larger markets. Bamboo offered a supplementary income opportunity for many households, but many poorer households remained poor and their gains from bamboo development were small.

The policies that facilitated the development of the processing sector also stimulated the demand for bamboo and improvements in production efficiency. They led to employment, income generation, and general development. Bamboo production and the bamboo-processing industry remained vital in the most developed counties, and many households continued to receive income or employment from their bamboo production activities. However, the industry lost its dynamic pioneering role as these counties developed and bamboo production became less important to overall countywide economic growth and welfare.

We can hypothesize that commercial forestry will follow a similar pattern if China further liberalizes its forestry markets. Farm households will take advantage of the opportunity. They will supplement their incomes and diversify their production, and some will eventually specialize in particular forest products. The greatest opportunities will occur where market access is good, and we can anticipate that China's immense effort in the 1990s to improve rural roads and other infrastructure will be important for this opportunity. However, we can also anticipate that the experience of China's poorest bamboo farmers will be duplicated as well. Poor forest households may improve their incomes, but these households are unlikely to be the primary beneficiaries of forest-based development, and many will remain poor even after forest development.

Environmental Impacts

China faces four serious environmental problems that are potentially related to trees, forests, and their uses:

- wind erosion and desertification,
- flooding and siltation of agricultural lands and the rural infrastructure,
- water pollution related to pulp and paper production, and
- the loss of natural biodiversity.

Wind erosion and desertification are a long-standing problem in northwestern and north central China, and China may be the world's greatest source of research on the topic. The causal physical links between forests and the other three problems are not as well known, the data are weak, and the serious literature about them often relies on thoughtful opinion rather than rigorous quantitative analysis. Moreover, we would argue that the authors of this literature and others who participate in the discussion often focus narrowly, overlooking the relationships among these four environmental problems. Enlightened policy is difficult, at best, under these conditions.

Sand and desert cover 2.5 million km^2 or 27% of China's land area, and desertification annually claims another 2,000 km^2 of landscape (Gluckman 2000). The government responded in 1979 with the Three North Forest Protection Project (Green Great Wall), an afforestation project that now extends to 24.3 million ha. The government has added several smaller afforestation and shelterbelt projects since 1979, voluntary citizen effort has added another 5.3 million ha, and many farmers have planted their own shelterbelts (Chapters 3 and 5). There is evidence of progress, yet desertification remains a paramount environmental problem.

Many viewed deforestation in the middle and upper watersheds of the Yangtze and Yellow Rivers as an important source of the vast flooding of the Yangtze and

Songhua watersheds in 1998. The central government responded with the Natural Forest Protection Program (NFPP), which bans logging in many of the upland watersheds leading to these rivers and restricts it in others. The largest share of state funds for rural environmental activities now supports forest protection, especially in these upland areas. However, it is not clear whether logging itself or general road construction in the uplands was more responsible for the flooding. It is clear that road closure in the harvest areas has been a successful means of controlling logging, but it is also clear that improved road construction in the mountainous NFPP regions would significantly improve erosion control.[9]

The pulp and paper industry is the largest source of rural environmental pollution, contributing 10% of China's industrial wastewater emissions and one-fourth of its chemical oxygen demand (State Council 1996). The government has experimented with a number of mechanisms for controlling these effluents, including pollution taxes, transferable pollution rights, and closure of more than 2,000 of the most offending small mills (Xu 1999). Nevertheless, the problem promises to worsen in at least one respect as a direct result of the logging ban associated with NFPP. Many pulp mills will have to increase their reliance on agricultural fibers as substitute inputs for wood fiber, but agricultural fibers are a greater source of both chemical and biological oxygen demand, and the pollution control technologies for pulp manufacture from agricultural fibers are not well developed.

Finally, the international environmental nongovernmental organizations have directed their attention to the losses of biodiversity that China must suffer as it draws down its stock of natural forests (discussed in Chapter 9). Do they overlook the new managed forests that have supported a 30% increase in China's total forest area since the beginning of reforms in 1978 (*China Forestry Yearbook* 1999)? Of course, these new managed forests are not as diverse as the natural forest, but they are not biological deserts either, and they surely support more biodiversity than the cropland and wasteland they replaced (e.g., Harkness 2001). Furthermore, managed forests are a substitute source for the extractive resources taken from natural forests. As such, the development of managed forests saves some amount of further destruction of natural forests—but how much? Our knowledge of the loss of biodiversity itself is limited, and we do not know the rates of substitution between managed and natural forests for either market or nonmarket forest products. Nor do we know the relative environmental gains obtained from converting wasteland and cropland into managed forest.

The task for China's environmental managers is to allocate their limited resources in a manner that yields the greatest net environmental improvement for all of China's society. This will not mean equal effort on all four of the serious environmental problems we identified. It will require improved physical data and a better understanding of the incentives motivating environmental improvement. Of course, China shares the problem of assessing tradeoffs among environmental objectives with all who would like to manage the environment better, but especially with the many developing countries where the environmental risks are so great, yet the environmental information is so limited.

Data and Analysis

This raises the final two points to be learned from China's experience, points about forest data and the analysis of many forest-related issues.

The analytical point is that two broad categories of forests, managed forests (including plantations and managed natural stands) and entirely unmanaged natural stands, must be separated for most meaningful analysis. The two categories respond differently to economic incentives, present different problems for monitoring and enforcement (M&E), and produce different environmental services. That is:

- Higher prices for forest products are an incentive for increased extraction from the natural forest, but they also encourage increased planting and other silvicultural activities that expand the area and volume of managed forests.
- M&E is difficult on the natural forests because they tend to be dispersed and remote. Trespass onto these properties is easy. Managed stands, in contrast, tend to occur in more accessible locations, and they tend to possess greater market value. The incentives for private M&E are greater, and any institution's M&E is generally easier on these managed forest properties.
- The more remote natural forests tend to provide greater biological diversity and a wider range of nonmarket environmental services (e.g., herbs for traditional medicines, mushrooms, and wildlife) than their managed counterparts. Indeed, the authors of Chapters 6 and 7 made this a central point in their discussions of general policy effects on forests, and the authors of Chapter 9 made it a central point in their discussion of biodiversity.

Close examination of China's forest data and personal examination of its forests call attention to three subsidiary points about forestry data.

- China's forest inventory includes as forests the land area in bamboo and also that in *economic forests* (horticulture, tea, and rubber). Bamboo and economic forests are important in China, but they are not part of the forest data for most other countries. China's forest data also include shelterbelts for protection against wind erosion, some of which are not dense enough to be included in the official forest inventories of many other countries.
- China changed its definition of *forest* from 30% forest cover to 20% forest cover in its most recent (1994–1998) forest inventory. (The minimal measure of standing forest volume necessary to define a forest did not change.)
- Very few of China's managed stands are intensively managed forest plantations. Many *afforested* lands may be called plantations, but few are intensively managed. Indeed, this was one of Zhang's criticisms in Chapter 5. In any event, the newly afforested lands probably contain more diversity than in their former condition as cropland or wasteland. Most *reforestation* in China occurs as interplanting within the existing degraded forest that remains after higher-valued trees have been removed. Foresters object to this as *high-grading*, a practice that leads to poorer quality future stands of timber. Undoubtedly it alters local biodiversity, but we have no evidence that it decreases overall diversity

and the excess of tree species of lower commercial quality as well as gnarly older trees may create a more favorable habitat for some species.

The lessons for China and for forest policy analysis worldwide are, first, that managed and natural stands must be separated in most analyses and in the policies designed for them. Commercial timber production from one reduces the level of harvests taken from the other, but in most other respects the two are not substitutes, and economic incentives such as higher prices have opposite effects on the two, inducing investment in managed stands but harvesting in the natural forest. And, second, sweeping intertemporal and international cross-sectional comparisons of forest trends must remain suspect until the analyst explains the underlying forestry data and accounts for any differences in the basic measures of forest cover or forest volume. The standard measures defining China's forests changed in the mid-1990s, and they were never entirely consistent with the measures used in many other countries. Do we run into the same problems with the forest inventory data from other countries? We contend that these two points are crucial, that a very large share of the global analytical literature overlooks them, and that the oversights lead to errors in policy.

Conclusion: Emerging Issues after a Quarter Century of Reform

Along with China's rapid economic growth in general, both the government and the general public have increased the level of their attention to environmental issues. Meanwhile, the disparities in regional welfare have become more noticeable. Erosion control remains a critical environmental problem, but the focus has shifted from wind erosion and shelterbelts to deforestation in the uplands and its relationship to flooding in the lowlands. The importance of forests and parks has grown along with the population's financial means to enjoy them, and the government has set aside 76 million ha in natural reserves, most of it in the 1990s. The Gini coefficient describing income inequality has risen more than 0.15 to the neighborhood of 0.35–0.40 (Chai and Chai 1994, Chen et al. 1995). Development lags, especially in the rural areas of the West, and income inequality and the unequal sharing of the benefits of development have become important political issues.

In 1998, the central government responded to these emerging policy issues with two large and related long-term programs, the Western Regional Development Program (WRDP) and the NFPP. The WRDP emphasizes the development of infrastructure. New and improved road and railroad construction is in process nearly everywhere we have traveled in western China during the last four years. This program also contains an element known as "Grain for Green," which compensates farmers with grain and with contracts for long-term land-use rights in return for removing their livestock and planting trees on steep-sloping cropland and pastureland.

The NFPP contains two principle elements, a forest-environmental protection element and a related element of financial support for the state-owned forest bureaus. The program was initially approved for two years in 1998, then renewed through 2010 as a 96.5 billion yuan (US$12 billion) effort that is 80% centrally funded and 20% funded by the provinces. Seventeen provinces, municipalities, and autonomous regions participate in NFPP. Eleven are also part of the WRDP. The remaining six are in the North and Northeast, the traditional center of industrial forestry.

More recently, NFPP was incorporated into the WRDP. The two programs together plan to protect 61.1 million ha of ecologically valuable but fragile watershed in the upper Yangtze and the upper and middle Yellow River Basins, plus 33 million ha in the North and Northeast. Approximately half of the western area will be closed for natural regeneration. The second half is on sloping croplands where the Grain for Green program provides the incentive. Planned harvests in the West had already declined from 13.51 million m^3 in 1997 to 1.128 million m^3 by 2000. The smaller level of continuing harvests is reserved for local use. Planned harvests in northern China declined from 18.54 million m^3 in 1997 to 12.13 million m^3 in 2001. A total of 8.6 million ha will be afforested, of which 1.635 million ha had been afforested by July 2001. 740,000 employees of SOFEs in the Northeast expect to change their employment or become unemployed (Hyde and Liu 2001).

These are young programs and their eventual impacts are uncertain. The intended impacts are, first, reduced pressure on the forests and reduced flooding downstream and, second, reorientation and recovery of many SOFEs, but on a smaller scale. There is already some evidence of forest recovery, and farmers welcome the Grain for Green program because its grain payments are often larger than the farmers' yields would have been from the recovered sloping land. However, the relationship between forest closure and downstream flooding is not clear, and the effects on local communities are unexamined—including the expected losses of income and employment from activities related to commercial forestry and the losses in terms of forgone opportunity to graze livestock on the newly closed forest. Furthermore, it is also unclear whether many small towns and villages can absorb the former employees of the SOFEs and provide the social services (hospitals, schools, etc.) that the SOFEs provided their employees. Will general economic growth in the West and the Northeast be sufficient counter these large changes and absorb the newly unemployed?

The uncertainty about the eventual impacts of these programs is great, and it is bound to induce short-sighted private economic decisions, as well as longer-term economic hardship for some. In our opinion, the central and provincial governments can reduce some of the uncertainty by permitting the local SOFE managers greater discretion to adjust their own operations in preparation for downsizing in 2010 and by preventing overzealous local administrators from applying NFPP restrictions (intended for state lands) to household and community forest operations.

The best time for adjustments in the operations of the SOFEs is when prices and interest rates are both low. Economic harvest criteria and the current government regulations on timber harvests are most comparable under these conditions. Fortunately, these are conditions that currently prevail (Xu 2002). Will the government take advantage of this opportunity to transfer some SOFE responsibilities to their employees in a manner similar to the transfers of collective lands and forests that occurred under the household and contract responsibility systems between 1978 and the mid-1980s? The government is considering this option, and the idea bears watching.

Meanwhile, if private sector forestry is going to enjoy the growth China has come to expect, a growth experience that can help offset some of the inevitable negative local community effects of NFPP, then the government must learn from some of the lessons outlined in this chapter. In particular, private forestry still confronts large elements of policy uncertainty and financial disincentive, and these must be corrected.

The government can reduce the uncertainty by eliminating its restrictions on private timber harvests and shipments. In fact, it should anticipate that the improved infrastructure in western China will decrease hauling costs, permit more distant shipment, and enable processing facilities like pulp mills to take advantage of their great economies of scale.[10] Continuing the government's restrictions on shipping permits will limit these cost advantages—and also limit the technological advantages that larger wood-using pulp and paper operations have in pollution control.

Tax reduction is another key to improved financial incentives. Taxes and other charges on private forest production absorb 35–60% of total revenues. (They are so high that, in some cases, decreasing these fees may decrease the incentives for tax avoidance and even increase the government's revenue recovery.) In comparison, agricultural taxes and charges account for about 10–12% of agricultural revenues, and the fees on bamboo run a little over 20% of revenues from the sale of bamboo culms (personal communications, Jikun Huang and Manuel Ruiz Pérez, February 19, 2002). It is not apparent that the forestry sector receives government services sufficient to account for the differences. The government is aware of the problem, and the State Forestry Administration and several provinces are experimenting with programs to consolidate and reduce forest taxes (Chapter 3). Nevertheless, many county tax collection authorities resist, because forest taxes and fees provide the financial support for their administrative staffs.

Decreasing the uncertainty and improving the financial incentives will improve the general climate for private domestic investment in forestry. It will also improve the climate for foreign investment—which plays an increasing role following China's new membership in the World Trade Organization (WTO). And just what will be the impacts of WTO? WTO will probably have a greater impact on the more capital-intensive pulp and paper industry than on any other component of the forest sector. Will it encourage the scale economies we previously indicated? Will it induce general growth and important spillover effects on forestry and the rural poor? Finally, if world markets in carbon develop, then we can anticipate that many marginal forest investments will become productive. We will speculate that all of the opportunities for new investments will work together to increase forest cover and standing forest volume if China can successfully address the difficulties involved with government control of the market, including its problems with high taxes, an uncertain policy environment for farm forestry, and financial support and operational scale in the SOFEs.

In sum, it is clear that the central government has the ability to initiate immense projects such as NFPP and Grain for Green. However, the costs of these projects are great—perhaps more than US$12 billion, an amount that far surpasses the historical contribution of logging from the natural forest. Furthermore, neither the institutional arrangements nor the incentives for achieving the desired results have anticipated either the economic efficiency of NFPP or the impacts of the long-run human adjustments that it will require. Moreover, despite its success initiating large projects, the government has not been successful in addressing issues related to timber pricing and taxation in the South. The fate of forests in the South has implications for the economy and the environment that are just as great as those of NFPP for the West and the Northeast. The South is a major region for forestry now, and its forest resource will only become more important if the NFPP restrictions on the forests of the Northeast and the West are effective.

These factors, taken altogether, mean that forest policy and management in China will continue to attract widespread interest, and they are a strong argument for thoughtful assessment before venturing very far into any of the new programs. Undoubtedly there will be new lessons to be learned. We think this is an argument to develop a system of sound, long-term policy analysis, a system designed to assess and recommend consistent forest policies that will be beneficial to all of Chinese society.

Notes

1. Perkins (1990) and McMillan and Naughton (1992) provide concise statements of China's experience..

2. The experience of agricultural reforms is well documented. See Lin 1992 and Chai 1997 for good reviews.

3. The summary data are incomplete for the two (of 10) provinces in this region with the smallest aggregate forest cover.

4. The advantages of the shareholding systems are unclear. Yin (1998) argued that the shift to shareholding systems was only a name change that resulted in no significant advantage over centralized management by the collective.

5. Most CRS rights had been renewed by the late 1990s, at the end of the first contract period, and transfers of these rights to other households received official sanction in 1995. Both the contract renewals and the improvements in rights represented by the sanction of transfers added to household confidence in the durability of existing forest land-use rights.

6. These risks are not inherent in forestry. For example, investments in forest management companies are *less* risky than investments in other firms and industries trading in U.S. stock markets (Sun and Zhang 2001).

7. Rozelle et al. also examined effects on standing forest volume. Their findings for the determinants of forest cover and forest volume were roughly similar.

8. The area in natural forest declined with general economic growth before 1985, although the response to general economic growth was inelastic and not nearly as emphatic as the opposite increase in managed forest area. The remaining natural forests in Hainan were legally preserved in 1985, and the forest inventories conducted since then suggest that economic growth since 1985 has occurred without any further loss of natural forest cover.

9. The contribution of road construction to flooding was an important item of discussion in the May 2001 annual meeting of the Forest and Grassland Task Force of China's Council for International Cooperation on Environment and Development.

10. Papermills in North American and Europe range from several hundred thousand to two million tons of annual capacity. In contrast, the capacity of the largest mill in China is only 250,000 tons and any mill over 30,000 tons is classified as "large." China has more than 1,000 state-owned and collective mills, and nine times as many other (private or township and village enterprise) operations, most with capacities of less than 5,000 tons (CTAPI 1993). The potential for cost saving through consolidation and the expansion of scale is immense.

References

CFIC (China Forestry Information Center). 1998. *Development of China's Wood Industry.* Beijing: Chinese Academy of Forestry.

Chai, Joseph. 1997. *China: Transition to a Market Economy.* Oxford, U.K.: Oxford University Press.

Chai, Joseph C.H., and Karin B. Chai. 1994. Economic Reforms and Inequality in China. *Revista Internationale di Scienze Economiche e Commerciali* 41(8): 675–696.

Chen, Shao-hua, Gaurav Datt, and Martin Ravallion. 1995. *Is Poverty Increasing in the Developing World?* Policy Research Department, World Bank, Data Appendix, updated version. 40: 359–76.

China Customs Office. 1999 through 2001. *China Customs Office Statistical Yearbook.* Beijing: China Customs Office Press.

China Agricultural Yearbook. 1985. [In Chinese]. Beijing: China Agricultural Press.

China Forestry Yearbook. 1987, 1989, 1998, 1999. [In Chinese]. Beijing: China Forestry Press.

Chomitz, Kenneth, and David Gray. 1996. Roads, Land Use, and Deforestation: A Spacial Model Applied to Belize. *World Bank Economic Review* 10(3): 487–512.

CTAPI (China Technical Association of Paper Industry). 1993. *Almanac of China's Paper Industry* [in Chinese]. Beijing: China Light Industry Publishing House.

Deacon, Robert T. 1994. Deforestation and the Rule of Law. *Land Economics* 70: 414–430.

FAO (Food and Agriculture Organization of the United Nations). 2001. *Global Forest Resources Assessment 2000.* Forestry paper 140. Rome: FAO.

Gluckman, R. 2000. The Desert Storm. *Asiaweek* October 13, 36–40.

Harkness, James. 2001. Introduction. In *WWF China Annual Report 2000.* Beijing: World Wide Fund for Nature–China.

Hirschman, Albert O. 1978. *The Strategy of Economic Development.* New York: Norton.

Hyde, William, and Jinlong Liu. 2001. NFPP and Its Effect on the Forest Sector. Unpublished report completed for the Chinese Academy of Forestry (under United Nations Development Programme project CPR/00/121).

Jepson, P., J.K. Jarvie, K. MacKinnon, and K.A. Monk. 2001. The End for Indonesia's Lowland Forests? *Science* 292: 859–861.

Li, L. 1996. The Causes of the Loss of Public Assets in Forest Industry Enterprises. *Forest Resources Management* 4: 47–49.

Lin, Justin Yifu. 1992. Rural Reforms and Agricultural Growth in China. *American Economic Review* 81: 34–51.

Liu, Dachang. 2001. Tenure and Management of Non-State Forests in China since 1950: A Historical Review. *Environmental History* 6(2): 239–263.

Lopez, Ramon 1998. The Tragedy of the Commons in Cote d'Ivoire Agriculture: Empirical Evidence and Implications for Evaluating Trade Policies. *World Bank Economic Review* 12(1): 105–132.

McMillan, J., and B. Naughton. 1992. How to Reform a Planned Economy: Lessons from China. *Oxford Review of Economic Policy*. 8: 130–143.

Pandey, K., and David Wheeler. 2001. Structural Adjustment and Forest Resources: The Impact of World Bank Operations. Policy research working paper. Washington, DC: World Bank.

Perkins, Dwight. 1990. Completing China's Move to the Market. *Journal of Economic Perspectives* 8(2): 23–46.

Ruiz Pérez, M., M. Fu, X. Yang, and B. Belcher. 2001. Towards a More Environmentally Friendly Bamboo Forestry in China. *Journal of Forestry* 99(7): 14–20.

Singh, Hemut. 1994. The Green Revolution in Punjab: The Multiple Dividend, Prosperity, Reforestation and the Lack of Rural Out-Migration. Unpublished paper. Cambridge, MA: Harvard University, JFK School of Public Policy.

State Council. 1996. Resolution on several issues in environmental protection [in Chinese]. Unpublished document no. 37.

Sun, C., and Daowei Zhang. 2001. Assessing the Financial Performance of Forestry-Related Investment Vehicles: Capital Asset Pricing Model vs. Arbitrage Pricing Theory. *American Journal of Agricultural Economics* 83(3): 617–628.

Westoby, Jack. 1987. *The Purpose of Forests: Follies of Development*. Oxford, U.K.: Basil Blackwell.

Wunder, Sven. 2001. Poverty Alleviation and Tropical Forests: What Scope for Synergies. *World Development* 29(11): 1817–1833.

Xu, Jintao. 1999. China's Paper Industry: Growth and Environmental Policy during Economic Reform. Ph.D. thesis. Blacksburg: Virginia Polytechnic Institute and State University.

Xu, Jintao. 2002. Economic Analysis of Logging Quota System in China. Paper presented at the International Forum on Chinese Forestry Policy, Beijing, June 13–14 (available from China's Center for Agriculture Policy Research, Chinese Academy of Sciences).

Yin, Runsheng. 1998. Forestry and the Environment in China: The Current Situation and Strategic Choices. *World Development* 26(12): 2153–2167.

Young, Carlos. 1996. Economic Adjustment Policies and the Environment: The Case of Brazil. Ph.D. thesis. London: University of London Department of Economics.

Zhang, Yaoqi. 2001. Economic Transition and Forestry Development. Ph.D. thesis. Helsinki: University of Helsinki Department of Forest Economics.

Zhong, M.G., C. Xian, and Y.M. Li. 1991. A Study of Forestry Development in the Major Agricultural Areas. In *Studies on China's Forestry Development*, edited by W.T. Yong. Beijing: China Forestry Press.

Index

For Product Safety Concerns and Information please contact our EU representative GPSR@taylorandfrancis.com
Taylor & Francis Verlag GmbH, Kaufingerstraße 24, 80331 München, Germany

www.ingramcontent.com/pod-product-compliance
Ingram Content Group UK Ltd.
Pitfield, Milton Keynes, MK11 3LW, UK
UKHW022132190926
475702UK00013B/1763
* 9 7 8 1 1 3 8 9 4 6 2 7 9 *